T0192933

THE CALF WITH TWO HEADS

THE CALF WITH TWO HEADS

Transatlantic Natural History in the Canadas

Louisa Blair

Baraka
Books

Montréal

ISBN 978-1-77186-330-8 pbk; 978-1-77186-335-3 epub; 978-1-77186-336-0 pdf

Cover and Book Design by Folio infographie
Editing and proofreading: Blossom Thom, Anne Marie Marko, Robin Philpot

Legal Deposit, 4th quarter 2023
Bibliothèque et Archives nationales du Québec
Library and Archives Canada

Published by Baraka Books of Montreal
Printed and bound in Quebec

Trade Distribution & Returns
Canada – UTP Distribution: UTPdistribution.com

United States
Independent Publishers Group: IPGbook.com

We acknowledge the support from the Société de développement des entreprises culturelles (SODEC) and the Government of Quebec tax credit for book publishing administered by SODEC.

The Two-Headed Calf

Tomorrow when the farm boys find this
freak of nature, they will wrap his body
in newspaper and carry him to the museum.

But tonight he is alive and in the north
field with his mother. It is a perfect
summer evening: the moon rising over
the orchard, the wind in the grass. And
as he stares into the sky, there are
twice as many stars as usual.

- Laura Gilpin

Table of Contents

Acknowledgements

Thanks to my mother, who taught me to look at the natural world; to Anouk Gingras, who inspired this book; to my friend, mentor and editor Bob Chodos; and to Jean-François Gauvin, who gave me invaluable historical, scientific and editorial advice.

Foreword

My friend and I have always argued about the value of knowing the names of flowers, trees and birds. Why can't you just enjoy them without knowing their names? he asks. And I say if I know their names, I can understand and enjoy them more. Finnish botanist Carl Linnaeus agreed with me. "If you do not know the names," he wrote in 1751, "your knowledge of things also dies."

The more I know their names and their particularities, the more I notice; and the more I notice, the more deeply I am absorbed and lose myself in my true home, the earth.

Now that we are losing the earth as we have known it through climate change, I am taking pleasure in the natural world all the more, with the added poignancy of anticipated nostalgia. Otherwise there is only despair.

I did not grow up in a family of scientists, but we learned about the flowers, trees, birds and animals that we saw. We took guidebooks with us on trips, and consulted them when we came home from walks. We were lazier about identifying insects and shells and rocks, but we still examined them closely, marvelling over their different colours and shapes and textures. For me natural history and art have always been closely intertwined. As my mother is an artist, close observation of this kind was part of our upbringing. And like many of the people in this book, we collect stuff: our house is full of wildflowers, rocks, feathers, shells, driftwood.

Orchid, Miriam P. Blair, 2019

Although born in Canada, I grew up in rural Britain, and spent a lot of my youth either up trees or among them, wandering for hours along streams and through woodlands and fields by myself or with my sister. We knew our patch of the earth better than we knew our own thoughts.

I didn't think I could ever love anywhere as much—anywhere that didn't have the huge spreading English oaks, or beeches so old and wide that two people couldn't begin to ring them with their arms. Anywhere without hedgehogs whiffling in the brambles, or bluetits flying down to drink from the milk bottles left by our door. That unique quality of light on a misty English morning, the smell of damp moss in a deep twisty lane.

But when I crossed back over the Atlantic to Quebec, I came to love our maples that turn in the fall, our sharp-leaved red oaks, and the majestic white pines just as much. Some of the trees and birds and flowers were the same, some were slightly different, some were completely unfamiliar. I learned their names all over again. I know and love our Quebec patch of ground now, with its noisy aspen stands, devil's paintbrush blooming in the ditch, chickadees dee-deeing in the maples, and the snowy owl standing motionless in a field.

This experience of transatlantic re-rooting led me to inquire about others who had travelled from England or France to the Canadas (Upper and Lower Canada, now Ontario and Quebec), two hundred and more years before me, with a similar interest in observing, discovering and collecting the natural world.

First, I discovered that most of these naturalists were passionate amateurs like me. They too took great pleasure in exploring and naming and collecting. The word "scientist" was only coined in 1833.

Second, I discovered that this transatlantic re-rooting resulted in a whole new knowledge of natural history, and not just due to the differences in species. It was a knowledge co-constructed between Europeans and Indigenous people from the two sides of the Atlantic. The settlers who then made Canada their home, such as my ancestors, constituted a third party who began to negotiate this co-construction with Indigenous peoples on the one hand (knowledge about nutrition, pharmacology and topography, for example) and their colonial counterparts on the other (knowledge about taxonomies, theories and instruments). With their different ways of relating to nature and the shifting power dynamics, this complex triangular web of relationships brought about new understandings of the world.

Our understandings have been similarly influenced by the shifting relations between professionals, trained geologists for example, and amateurs, including members of local natural history societies, recreational gardeners, and walkers like myself. Natural history activities moved from field observations, drawing and collecting, museums and classifying activities, into academia-controlled laboratory-based teaching and research. The emphasis is now swinging back to field research, to "citizen scientists," or amateurs who throughout the world are providing an unprecedented quantity of observation-based data, and to an acknowledgement that Indigenous ways of knowing the earth are essential to our way forward.

We need to continue to carefully observe organisms in their environment, and to cross boundaries and disciplines—and not just the scientific disciplines—to understand them properly. We need to do this for our physical survival, e.g., for medicine, food security, conservation and landscape restoration, but also for our continuing joy. Art, history, and culture, as well as science, are all essential for understanding and loving the natural world enough to save it.

What really propelled me to become a botanist is that I wanted to know why golden rod and asters looked beautiful together. But then, when I went to study botany at the university, they told me very quickly that that was not a science question. That was an art question.

... that was very formative for me in thinking about these two ways of knowing, in knowing about both traditional ecological knowledge and the Indigenous perspective that I'd been raised in... I actually think that that goldenrod and asters experience has been one of the most profound ones in my profession as a naturalist and a professor, in trying to reconcile those two worlds.

- Robin Wall Kimmerer, author of *Braiding Sweetgrass*

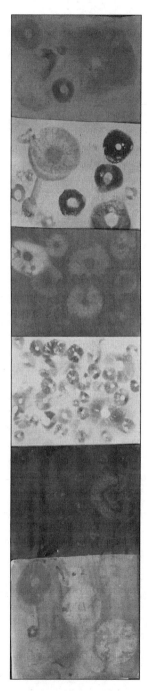

Mushroom prints. Miriam P. Blair, 2021

Introduction

This book explores the transatlantic passion for natural history that gripped travellers and settlers alike in the eighteenth and nineteenth centuries, and how this passion changed our understanding of the world.

With the colonization of the Canadas by Europeans, a new crop of explorer-naturalists set out to map and investigate the land, describing and trying to categorize the rocks, birds and plants they found. Animals, vegetables and minerals (a division established by eighteenth-century Swedish botanist Carl Linnaeus) were different in some ways, and yet the same in others, from those in Europe. The comparisons challenged existing classification systems, many of which assumed that species were constant through space and time. The discovery of "new" species elsewhere and of fossils, which suggested some species had disappeared, set off a debate, not yet over, on the origins of the world. Did the Biblical account of the beginning of the world skip some earlier episodes? Was Noah's Ark big enough? Could we really be descended from apes, rather than from Adam and Eve?

In reflecting on these findings, transatlantic naturalists were inspired by French and

Exit from Noah's Ark. Paris, Bedford Group, ca. 1420

British precedents, challenged by the theory of evolution by natural selection, and informed by the millennia-old knowledge of the original naturalists of the continent, the Indigenous peoples whose land they were investigating. Indigenous people showed them around, taught them how to travel, gave them maps, showed them plants and animals, and cured them when they got sick.

People from many walks of life, or amateurs, developed a passion for collecting, painting, describing, classifying and cataloguing the natural world with the goal of explaining and accounting for every corner of it. The word amateur comes from the Latin *amator,* a lover, or enthusiastic admirer. But recognized scientists were not always impressed with this onrush of enthusiastic amateurs trying to classify specimens. "The word *naturalist,*" wrote English biologist Thomas Huxley in 1859, "unfortunately includes a far lower order of men than chemist, physicist or mathematician ... every fool who can make a bad species and worse genera is a 'Naturalist'!"

Before photography, artists played a large part in these efforts to understand and communicate the natural world. To illustrate the new information that was reaching the public, artists in the field had to depict specimens as accurately as possible in quick sketches before their subjects rotted or were eaten by insects. Many of the greatest natural history illustrators were women.

Naturalists not only described and drew what they observed. Like children on a beach, they collected oddities and brought them home to add to their collections. The *cabinets de curiosités* of the sixteenth century, which contained such wonders of nature as a calf with two heads, or a unicorn horn, morphed into the first museums of nature. These led to universal exhibitions, and finally to the turn that natural history has taken today.

I – NOVELTIES OF NATURE

The study of natural history, or the observation and description of natural objects and their links with the environment, reaches back at least as far as Aristotle and ancient China. With transatlantic travel, knowledge of natural history expanded in the fields of astronomy (stars), botany (plants), zoology (animals), entomology (insects), ornithology (birds), geology (rocks) and palaeontology (fossils). A brief look at the changes in just four of these gives a glimpse of new forms of knowledge that resulted from these transatlantic adventures.

The heresy of the stars

Astronomical leaps

Accurate charts were necessary for transatlantic travel, and astronomy, one of the oldest sciences, was necessary for making those accurate charts. Knowledge of astronomy had undergone a universal expansion during the seventeenth and eighteenth centuries.

Early freelance explorers such as Vasco da Gama, John Cabot and Ferdinand Magellan, who were looking for cheaper cooking spices, had to make do with the most rudimentary of charts, sailing their ships into a blank space where the cartographer's knowledge ended. The earliest maps were mythical drawings that often depicted equally mythical creatures. French mapmaker Pierre Desceliers, who illustrated Jacques Cartier's voyages, showed unicorns in James Bay, men with dogs' heads in the Hudson Bay, and in the Moisie River, a horse-shaped fish that came ashore at night. Writer François Rabelais made cruel fun of Cartier's stories, but their audience back home in France was more interested in these fantastical accounts than in accuracy. When Samuel de Champlain arrived in New France in the 1600s, he brought a new spirit of rigour and objectivity. His 1632 map,

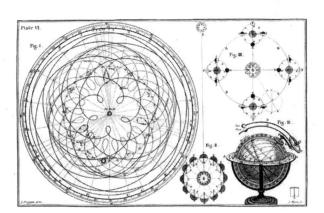

The planetary motions as seen from the earth. James Ferguson, "Astronomy Explained upon Sir Isaac Newton's Principles," 1756

North American owls, by Alexander Wilson. *Edinburgh Journal of Natural History and of the Physical Sciences*, Vol. 1, 1835.

informed partly by Indigenous people, was the foundation of all other maps of Canada for many years.

Early astronomical science for determining longitude, or one's east-west position on the earth, consisted of timing lunar eclipses in different parts of the world simultaneously. The time difference in different places would be proportional to the difference in geographical longitude by 15 degrees per hour.

A lunar eclipse saved Christopher Columbus' life in 1604. He and his worm-eaten, leaking ships were stranded on the island of Jamaica, where the host population was getting tired of supporting them. He warned them that the moon would disappear if they did not cooperate, and the next night there was a total lunar eclipse, of which he was perfectly aware from reading his almanac. They begged him to restore their moon,

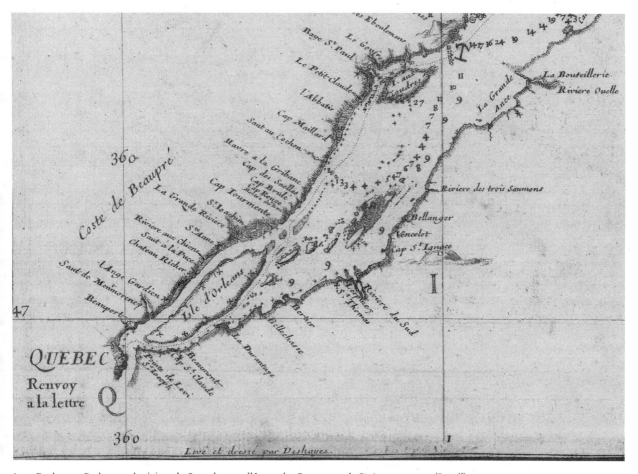

Jean Deshayes, *De la grande rivière de Canada appellée par les Europeens de St. Laurens*, 1715 (Detail)

and when it reappeared, he managed to negotiate food and protection for his troops until the arrival of a supply ship.

Useful as it was, the timing of eclipses was limited to observations on land, as clocks did not work at sea. The most accurate clocks available to early navigators were water clocks and sand clocks such as the hourglass. To measure how fast he was going, Columbus threw a piece of driftwood overboard ("the ship log") and timed how long it took to pass him by: his speed was the length of the vessel over that time. The origin of the word *knot*, meaning nautical miles per hour, comes from a sixteenth-century sophistication of this technique consisting of attaching a rope to the ship log with knots tied at regular intervals. Using an hourglass, one sailor would time how long it took for the knots to pass through the hands of another as the rope rolled overboard. Columbus tried navigating by the stars, but failed, returning time and again to his magnetic compasses and his hourglass.

In 1714, the British offered £20,000 to anyone who could develop a clock that worked at sea. A self-taught clock maker, a carpenter named John Harrison, won most of the money fifty years later after bringing a succession of maritime clocks to the Longitude Board.

In New France, the Jesuits had been teaching navigation since 1661, as pilots able to navigate the St. Lawrence River were a vital necessity. Marine surveyor Jean Deshayes's timing of a lunar eclipse in Quebec in 1685 had helped the French establish the exact longitude of the city, and his maps were used by pilots until the final years of the French regime.

However, at Quebec's Collège des Jésuites, the professors' celestial calculations were obscured by the fact that they were still not allowed to teach Copernicus' heretical idea, published more than a century earlier, that the earth revolved around the sun rather than the other way around. Finally in 1751 they began to teach that Copernicus' theory was right and not just a hypothesis—although Rome took another 70 years to revoke its condemnation of the idea. With the closing of the Collège des Jésuites in 1759, two thousand years of Aristotelian cosmology crumbled into stardust.

A hardier mulberry

Botanical usefulness

With the discovery of the Americas, Europeans hastened to collect the plants they found, both for their usefulness and to see how they fit into—or defied—contemporary Linnaean classification boundaries. Kings sent botanists to bring back specimens. In 1697, France commissioned Michel Sarrazin, who for 20 years sent specimens back from New France to the Académie royale des sciences. He picked plants, noted characteristics of their habitat, geographical location and Indigenous knowledge about them, and sent them home on a ship. To his chagrin, he was never made a member of the Académie, but was given the dubious honour of having a carnivorous plant, *Sarracenia*, named after him.

Much of the plant trade was for money. The Haudenosaunee (Iroquois) showed the Jesuit priest Joseph-François Lafitau a kind of ginseng, which in the early 1700s sold in China for 60 times the Montreal price. Sweden sent botanist Pehr Kalm to North America, including to Quebec in 1748, to look for plants of use to agriculture and industry. The governor of New France, Barrin de La Galissonnière, a keen naturalist himself, instructed his troops to help Kalm collect specimens, and promised to promote the most zealous collectors. Kalm was looking for hardy mulberry trees for the Swedish silk industry, and hoped the sugar maple tree and the fast-ripening maize he found in New France could be domesticated in Scandinavia.

The British, for their part, expected William Jackson Hooker, Director of Kew Gardens, to identify new plants that could assist British and colonial farmers and gardeners. Hooker chafed under this directive. He was more interested in taxonomic and biogeographical research. In addition, botany was a hugely popular source of pleasure for his visitors to Kew Gardens. By the mid-1870s, Kew welcomed almost 700,000 visitors a year, and 58,000 on one day alone in 1877. To remind visitors of their colonial connections, the Kew flagpoles were made of gigantic Douglas firs from Canada.

Pet pigeons

Ornithological obsessions

The romance with nature focused perhaps even more intensely on birds. The Comte de Buffon's *Histoire naturelle des oiseaux* (1770-1783) was based on the French royal ornithological collections, the most comprehensive and exotic in the world. It was a European best seller. People classified birds, catalogued, drew, killed and stuffed them, or caged them and bred them. Charles Darwin himself kept 90 pigeons, bred them and then observed their variations to establish that they came from a common ancestor, the rock dove.

Birds were also fashionable in another sense. Women bought hummingbirds, unknown in Europe, for tuppence each to pin to their hats. One Parisian milliner used 40,000 hummingbirds per season for this purpose. In 1889, a group of British women established the Plumage League, later the Royal Society for the Protection of Birds. They campaigned against the fashion of wearing of feathers in hats, and recruited famous writers and eventually even Queen Victoria herself to the cause. Across the ocean in the city of Quebec, naturalist Harriet Campbell Sheppard published the first report on birds in the city, and

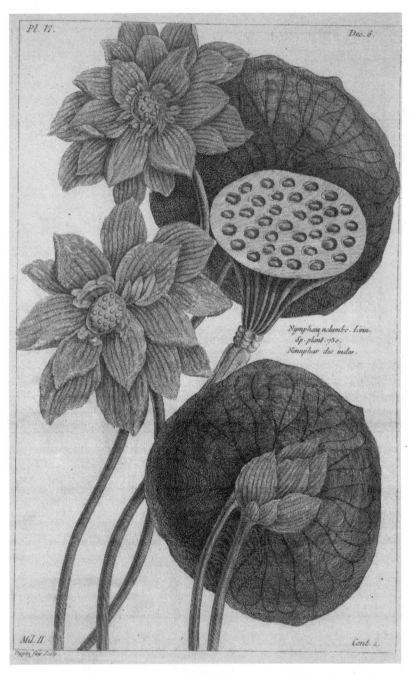

Water lilies, engraved by Jean-Victor Dupin, in Buchoz, P-J., *Histoire universelle du règne végétal, ou nouveau dictionnaire physique et economique de toutes les plantes qui croissent sur la surface du globe*, 1775-1778

Kew Garden's flagpole felled in British Columbia by MacMillan Bloedel, Copper Canyon Division, 1958

historian William Wood enlisted his relatives in the British royal family in his efforts to establish Quebec's first bird sanctuaries on the North Shore and in Labrador.

No rotting flesh

Geological dating

Johann Charpentier had created a geological map of Saxony as early as 1778 for mining purposes, but it was William Smith's brightly coloured geological map of England (1815) that set off Britain's natural history craze for geology.

Most exciting was speculating on the age of the world by examining the different types of fossils contained in successive layers of rock (strata), a kind of natural archive

Helianthea osculans (buff-tailed star-frontlet hummingbirds), by William Matthew Hart, 1880.

of the earth's history. Strata were named after the places where they were most clearly visible—Jurassic in the Jura, Devonian in Devon—and then these names were given to periods. Until well into the eighteenth century, the fossils found in these rocks were not recognized as the petrified remains of living creatures. Once the penny dropped, the race was on to find out which fossils were found in which layers of rock, when these organisms had been alive, and why they no longer existed. If the Book of Genesis said the animals were created by God, did God also change them? What's more, Noah's Ark seemed to have missed a few of the animals. Could the Bible be wrong?

As people in Victorian Britain explored, surveyed, mined, quarried, or just went for walks in the countryside, they eagerly participated in this attempt to date the world. For the middle classes, palaeontology became a glamorous and genteel pastime, even for ladies. Fossilized creatures, after all, no longer smelt of rotting flesh or dripped bloody entrails.

While the British were finding sea monsters, the Russians and Americans were finding fossils that looked like giant elephant bones. The fossils suggested an exotic prehistoric world populated by species that had vanished. Various geologists proposed explanations for how these extinctions had occurred. In 1813, French comparative anatomist Georges Cuvier said there must have been a series of violent catastrophes before Genesis, while geologist Charles Lyell said in 1832 that the changes had been gradual, over hundreds of thousands of years. German botanist Anton Friedrich Spring had a philosophical explanation: he believed that although God's creative thought gives nature a kind of archetypal unity, change was intrinsic to each species: "Everything in nature is continuously becoming," he wrote in 1838. "Species do not persist in a state of 'being,' but instead are continuously 'becoming.'"

Few entirely dismissed the Genesis account of creation. Until the seventeenth century it was the only surviving textual record of ancient history known to most Europeans. As additional sources were found, geologists added the textual evidence to their theories. They gradually accepted that the dating of the Creation at 4004 years ago, as calculated by Irish historian James Ussher in 1650, could be quite wrong.

Whereas now Darwinism is claimed by atheists as a weapon against modern Biblical literalists, in the mid-nineteenth century there was little argument among Christians about the reality of massive evolutionary changes as shown by the fossil record. Many of the most avant-garde British geologists were clergymen, yet were happy to rethink Creation as happening over six epochs instead of six days. And the fossil record seemed to confirm that creation was indeed a linear story, as the Bible suggested. With Mary Anning's discovery of a 17-foot "fish-lizard" in Dorset, England, in 1823, it looked as though fish came first, and then reptiles.

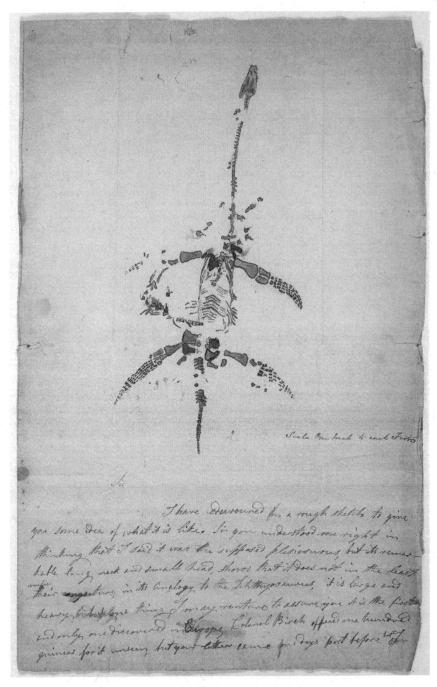

Sketch in a letter by Mary Anning of the plesiosaurus she found, 1823. "One thing I may venture to assure you it is the first and only one discovered in Europe."

The big question was, what caused these changes? And what was the place of humans in all this? If we were really descended from apes, as Darwin later proposed, what did that say about the "irreducible dignity of man"?

Before we laugh dismissively, theoretical physics today is still raising equally extraordinary questions. Could the planet disappear through a tiny black hole? Is there a Theory of Everything? The discovery of geological time, or "deep time," may have changed our perception of our place in the cosmos more than any other discovery.

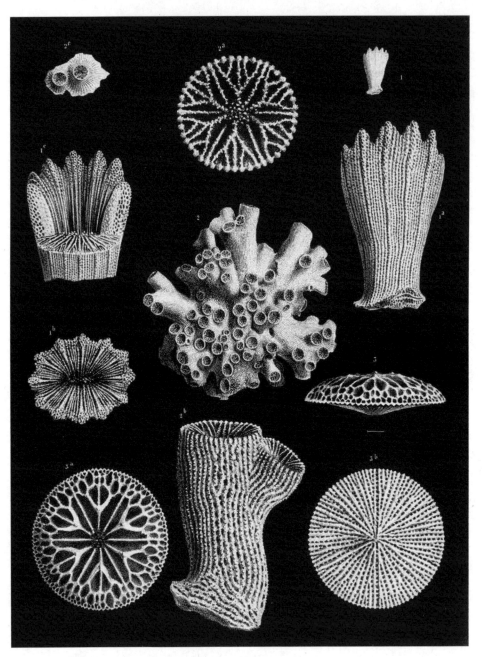

Coral fossils found in London clay, by J. Delarue, ca. 1850.

II – INTERROGATING NATURE

American author Barry Lopez has defined natural history as "the patient interrogation of a landscape" in reference to Inuit knowledge of nature. Unlike Europeans, Indigenous people did not consider themselves as separate from nature. They were aware that they directly depended on it for every aspect of their lives and, consequently, had a deep knowledge of the natural world and how to move around in it.

This chapter shines a few spotlights on natural history in Quebec, first as practised by some of its Indigenous inhabitants, and then by a group of transatlantic inhabitants of the city of Quebec in the early nineteenth-century.

Memory maps

Indigenous natural history

Before the arrival of Europeans, several hundred Indigenous nations occupied North America. They were organized into villages, bands and confederacies, and spoke 221 different languages. The natural world was teeming with birds, mammals and plant life. In what is known today as Quebec, Algonquian and Haudenosaunee peoples fished, used plants as food, medicine and dye, and grew crops. Inuit had millennia-old highly specialized skills for living in a uniquely difficult environment.

Map-making was already a widespread practice among Indigenous people when Europeans arrived. They drew ephemeral maps on the ground or in snow and memorized them. They also drew maps on bark or skin. Indigenous people informed the earliest European-made maps of the continent.

In 1604, six Indigenous travellers drew a map of the coast of Maine for Champlain in charcoal. When he sailed along the coast later, Champlain "recognized everything in this bay that the Indians at Island Cape had drawn for me." The Jesuit Paul Le Jeune recounts how an Innu (Montagnais) warrior, spontaneously taking a pencil in hand, depicted the country of the Iroquois enemies where he

> They draw the most exact maps imaginable … there's nothing wanting in them but the Longitude and Latitude of places. They set down the true north according to the Pole Star; the Ports Harbours, Rivers, Creeks and Coasts of the lakes; the roads, mountains, woods marshes, meadows etc. counting the distances by Journeys and half-journeys of the Warriors and allowing to every Journey Five leagues. These maps are drawn upon the rind of your Birch Tree…
>
> - Baron de Lahontan, *Mémoires de l'Amérique septentrionale*, 1703

The Haudenosaunee people making maple syrup, in *Mœurs des sauvages ameriquains comparées aux moeurs des premiers temps*, by Joseph François Lafitau, 1724.

was going to wage war. Another Jesuit, Louis Nicolas (see page 51) sang the praises of the navigational abilities of the Algonquin people he travelled with. They never got lost, he wrote, even in long voyages "of four or five hundred leagues" (2000–2500 km) through the forest. They followed plantains, which tended to grow on the trampled ground where other people had walked, and were constantly aware of their orientation through, for example, noting which side of the trees the moss grew (mostly on the north). Their maps added unfamiliar dimensions: Indigenous cartographers measured distance by the length of time it took to get somewhere, or the number of nights spent, i.e. the chronography as well as the topography. Some maps might not distinguish between a river and a portage: the route itself was what interested them rather than whether it had to be walked or paddled.

Inuit navigated on moving sea ice using "memory maps" of the stars, recounted down the generations in ancestral stories. They had their own names for many constellations. On land, where trails can be erased by blizzards several times a year, Inuit mark routes with *inuksugait* (rock cairns; plural *inuksugaq*), rocks or other objects placed along the trail; but most routes are memorized, as established by centuries-old tradition.

Europeans devoured all the Indigenous cartographical knowledge they could, for Indigenous knowledge was the principal source of information for their maps and for the detailed accounts of their voyages that they sent home to convince their bosses that the sponsorship of their travels was worth every penny.

Indigenous people knew the properties of over 1000 species of wild edible plant, including seaweeds, algae, fungus and ferns. The land was still teeming with wildlife. The Huron-Wendat, who lived near Lake Huron, also ate the seeds, flowers, leaves and roots of wild plants. They gathered edible berries, harvested maple and birch syrup, and made flour from the bark of the birch, oak and tulip poplar. They farmed as well as hunted, growing squash, pumpkins, sunflowers and corn (or maize). Some used rotten wood and ground-up horseshoe crab shell as fertilizer. Corn, 60% of the Huron-Wendat diet, was roasted and ground for a travel food called *sagamité*.

In Europe in the sixteenth century, yield for cereals was six parts harvested for one part sown, at best. But the corn cultivated for millenia by Indigenous people, as a result of careful selective breeding, yielded 200 for one.

The Huron-Wendat also made tools out of stone—scrapers, drills, arrowheads, clubs and axes. Stone tools were used to work wood and leather to make arrows, toboggans, bowls, puzzles, wooden armour, moccasins, snowshoes, and canoes big enough for six people. They made needles and spoons out of animal bones, and made spears, seines, hooks, lines and nets for fishing. They used copper for tools and jewellery.

Indigenous doctors combined spiritual practices, such as the use of sacred objects, rituals and prayers, with empirical knowledge of wild plants and human anatomy.

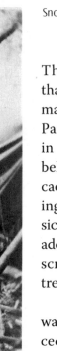

Snowshoe maker, ca. 1900-1925

The Huron-Wendat also believed that unsatisfied natural desires could make people ill. According to Jesuit Paul Rageneau, the Huron-Wendat in seventeenth-century New France believed that "one of the most efficacious remedies for rapidly restoring health is to grant the soul of the sick person these natural desires." In addition to song and dance, they prescribed teas, set broken bones and treated dislocated limbs.

Foxglove (*digitalis purpurea*) was used as a cardiac stimulant, and cedar bark (*thuya occidentalis*) to fight scurvy. A Haudenosaunee cure for scurvy famously saved the lives of Jacques Cartier's crew in Quebec in 1535. While some missionaries were reluctant to ask about Indigenous medicine as they did not want to encourage its spiritual aspects, other early settlers, explorers and missionaries took careful note of Indigenous remedies, and even made money out of them, such as the lucrative ginseng export to China. The sisters at Hôtel Dieu in Quebec began exporting maidenhair fern (*adiantum pedatum*) after learning that it could cure lung infections and help with childbirth. Pehr Kalm, Michel Sarrazin, Louis Nicolas and midwife Catherine Jérémie all learned about Indigenous herbal medicines and passed the knowledge on to Sweden or France.

This transatlantic sharing of knowledge involved injustices: Indigenous medicines and foods were commercialized without permission or compensation. The seizing of Indigenous harvesting grounds as well as forced language loss resulted in loss of Indigenous knowledge of the natural history of their territories. Most Indigenous communities have strict teachings that govern the harvesting of plants, fishing and hunting. Today, Indigenous knowledge of natural history and fraternal respect for the animals and plants that sustain us is essential for protecting and restoring the earth.

More fun than embroidery

Lady botanists of Quebec

In Quebec in the 1820s, three elegant women could often be seen wandering through the woods in Sillery, just past the Plains of Abraham in the city of Quebec. They were picking ferns, flowers (including native orchids), mushrooms and other wetland plants. This was not to decorate their dinner tables, although all three were famous for their magnificent dinner parties. Instead, they first tried to identify the plants. Then they sketched, pressed and dried them. They collected them in *herbaria* (books of pressed, dried and labelled plants) so they could be studied further. And if the women thought the plants very special, they were packed up and sent off on the next sailing ship to Britain.

All three women were passionate gardeners and important botanists before botany was a profession or an academic discipline, either of which would have excluded them. Lady Dalhousie and Anne Perceval ended up going back to Britain, from whence they had come. Harriet Sheppard was born and raised in Canada, and stayed.

Countess Dalhousie, or Christian Broun Ramsay, was the wife of Lord Dalhousie, the Governor General and founder of the Literary and Historical Society of Quebec (see page 38). She lived in the Château Saint-Louis, where the Château Frontenac is now, and as wife of a governor general had plenty of boring public duties to attend to. What she preferred was to ride off on her horse, Cherub, or drive out in a horse-drawn barouche with her friends to collect plants. Some of the plants, including a giant thistle, she sent back to her gardener at their castle in Scotland to adorn the flower beds that were specially reserved for North American plants. Others she slipped into her personal journal, which was bursting with dried plants and even with dried insects. Others again she sent to her friend William Jackson Hooker in Glasgow for his book *Flora Borealis-Americana* (1829-40), an imperial project to identify all the plants in British North America. In his essay "On the Botany of America," published in 1825, Hooker placed Lady Dalhousie in the "first rank" of people who were "industriously engaged in furthering the Flora of [Canada]."

> I have friends scattered about in every direction—Some I can exhort, some command, some entreat and some supplicate… Mr. and Mrs. Sheppard will take care of Quebec, Lady Dalhousie of Sorel & Montreal, and I of all they leave.
>
> - Letter from Anne Perceval to William Jackson Hooker, 1825

One of Lady Dalhousie's favourite fellow-botanizers was Harriet Campbell Sheppard, wife of a lumber merchant, who lived at Woodfield, a magnificent 100-acre estate just west of downtown Quebec. The Sheppards had a library of 3000 books, an art gallery, a natural history museum, several greenhouses and an aviary. Harriet Sheppard

Dalhousiea bracteata Grah., or Goopree (Sylheti name), plant renamed after Lady Dalhousie. By unnamed Indian artists, in William Roxburgh's *Plants of the Coast of Coromandel*, 1819.

contributed 144 entries to Hooker's *Flora Boreali-Americana* and was another botanist highly valued by Hooker and his colleagues in North America.

In 1829 she lectured at the Literary and Historical Society of Quebec on shells found in Quebec. Published in the Society's periodical, the *Transactions*, it is one of the earliest publications on Quebec conchology. She published again in the *Transactions* in 1835 on songbirds in Quebec, adding seven new ones to those already documented. She was the only woman ever to publish in the periodical.

The third of the trio was Anne Mary Flowers Perceval, wife of the collector of customs for Quebec. She lived next door at another magnificent estate, Spencer Wood. She is cited more than 150 times in Hooker's *Flora Boreali-Americana*, and plants she gathered can be found in collections in the US, Paris and London. She maintained a lively correspondence with eminent botanists worldwide, including American botanist John Torrey, who with Asa Gray wrote *A Flora of North America* (1838) and was one of the original members of the American National Academy of Science.

For William Jackson Hooker, Professor of Botany at Glasgow University and later Director of Kew Gardens in London, these women were a crucial source of local knowledge. He cited the three women 450 times. Hooker was a collector of collectors, cultivating thousands of colonial correspondents. His work depended on vast networks of personal contact and friendship, but also on people being willing to wander around in the woods in far-off lands.

Botany was considered a desirable feminine accomplishment for British-educated aristocratic women, and the trio devoured botany textbooks (Lady Dalhousie read seven in 1830 alone). "The study of Botany, that science by means of which we discriminate and distinguish one plant from another, is open to almost every curious mind," wrote the female author of a gardening book in 1799.

Portrait of Lady Dalhousie by Sir John Watson Gordon (1837). On the table beside her is a bird and a botanical drawing. Both bird (*Psarisomus dalhousiae*) and plant (*Dalhousiea bracteata*) were named after her.

For women who were married to colonial officers or merchants in British colonies, as was the case with Lady Dalhousie, Harriet Sheppard, and Anne Mary Perceval, there was an extra incentive. They could find an as-yet undocumented plant, which might be formally recognized back in Britain. They themselves might even be personally recognized—best of all was to have your name in Latin form bestowed on a plant you had discovered. Both Lady Dalhousie and Harriet Sheppard were immortalized in this way.

Although Hooker's collectors were mad to find new species, he preferred them to position plants on a continuum—"lumping" rather than "splitting." Hooker's son Joseph, who took over the directorship of Kew, complained that one Australian informant was "vomiting forth new genera & species with the lack of judgment of a steam dredging machine."

Collecting plants was satisfying and intellectually stimulating. It was an enjoyable way to explore the country with children and friends and add to one's garden. And in Quebec these women were perhaps more free to roam around, as well as to engage in intellectual activities, than they would have been back in restrictive Victorian Britain, where women in polite society had to sit at home, be decorative, and carry on with the needlework. In Charlotte Bronte's novel *Shirley* (1849), Reverend Helstone tells his ambitious niece, "Stick to the needle—learn shirt-making and gown-making, and pie-crust-making, and you'll be a clever woman some day."

By the middle of the nineteenth century, women were being increasingly excluded from public life in the Western world. The sciences, including botany, were no exception. By the 1850s, botany had been reshaped as a science for men.

A singular specimen of the potato

Natural history at the Literary and Historical Society of Quebec, 1824 to 1840

The kind of natural history that took place under the auspices of the Literary and Historical Society of Quebec was also an "interrogation of the landscape," but perhaps of a less patient kind than that practised by Indigenous naturalists. This institution, a learned society founded in 1824 and alive and well to this day, is a fascinating manifestation of the urgent passion for exploring, collecting and classifying that spread to the colonies.

The stated intention of the Society's founder Lord Dalhousie, Governor General of British North America at the time, was to collect and preserve documents about Canada's history, and especially about its Indigenous people. From 1829 to 1831, the Society published five papers on Indigenous topics, and acquired a Montagnais dictionary (1726) and a Huron-Wendat grammar compiled by the Jesuit missionary Pierre Chaumonot in about 1673. It also acquired documents on New France containing descriptions of Indigenous peoples, such as accounts written by Pierre-François Xavier

The library of the Literary and Historical Society of Quebec, founded 1824.

de Charlevoix, Claude-Charles Bacqueville de La Potherie, the Baron de Lahontan, and the Society of Jesus (the Jesuits). This initial interest in publishing on Indigenous topics soon gave way to a focus on investigating nature. Forty years after its founding, a Society member gave a paper in which he lamented that it had failed in Dalhousie's mandate, and that Indigenous peoples were "perceptibly fading away before the influence of European civilization, either dying out under its uncongenial system, or losing their identity in the unequal amalgamation."

Whereas Dalhousie dreamed of building up a local knowledge base about Canadian history, his wife dreamed of the equivalent for natural history. She had collected 382 specimens of Canadian plants in a herbarium, and instead of sending them back to Britain as usual, she gave them to the Society, the first donation to its museum collection.

The naturalists at the Society were doctors, lawyers, military officers, merchants, government officials, teachers, surveyors and churchmen. Two-thirds were anglophone,

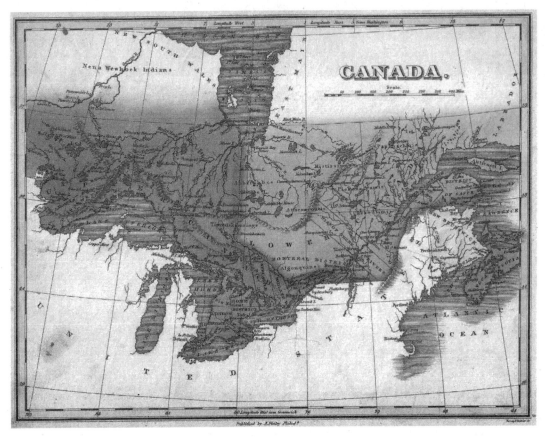

"Canada" in 1827. Anthony Finley's map names numerous First Nations, and includes Upper Canada (now Ontario), Lower Canada (now Quebec), New Brunswick, Nova Scotia and Prince Edward Island.

one-third francophone, and they were Jewish, Catholic and Protestant. They just had to be rich enough to afford the steep membership fees.

The House of Assembly in 1827 had voted funds for exploring the province of Lower Canada, and much of this mapping was done by members of the Society. Andrew and David Stuart went to explore the Saguenay, while Frederick Baddeley, a British military officer, went to the Baie des Chaleurs. With the goal of preparing for the settling of new immigrants and looking for natural resources, they surveyed the agricultural potential, geology and natural history of the country and took latitude and longitude measurements. Naval captain Henry Bayfield, sent by the British Admiralty, spent 14 years surveying the St. Lawrence River. They all brought specimens back to the Society, where they were classified, preserved and displayed in its museum. The museum filled up with minerals, wood, stone, fossils, shells, plants, birds and mammals, as well as Indigenous and other historical artefacts. They lectured about their adventures, and shared information about the customs and languages of the Indigenous people they had met.

The Literary and Historical Society became a kind of informal university of natural history, with lectures, classes, a science library, a publishing house, and the natural history museum. Such museums were seen as essential for learning about the natural world. Instead of the former practice of studying the classics (such as Aristotle's *Metaphysics*) to understand the workings of nature, members were building a new kind of knowledge based on detailed observations made in the woods, mountains and rivers, on rocks, minerals, fossils, weather, fish, crickets, rattlesnakes, bears, plants, birds, shells and stars. They used the Society's library and museum as a laboratory in which to study and display their findings.

This was before anyone had ever heard of plate tectonics, the Canadian Shield or an ice age. They travelled mostly by horseback, foot or canoe. They were keen observers and mostly excellent writers, and their observations were published in the *Transactions* and distributed worldwide. In return, the Society received periodicals from dozens of other learned societies around the world.

Such periodicals were the main media through which scientific thought evolved. Unlike books, periodicals could track changing cultural norms and the latest theory could be debated from one issue to the next. Published from 1829 to 1924, the *Transactions* provided the first serious and consistent outlet for scientific publication in Canada. Until 1843, 76% of papers published in the *Transactions* were on natural history. On the subject of geology, in particular, the Society held a virtual monopoly on publication in Canada until 1836.

The Society offered a variety of other ways to learn about natural history, too. Its library contained the most important scientific books available at the time, in both French and English, including Michael Faraday's *Chemical Manipulation* (1827) and Georges Cuvier's *Le règne animal* (1816).

The Society collected scientific instruments such as globes, telescopes, barometers, and hygrometers which could be used for lectures and public demonstrations, or borrowed by members for research purposes. Frederick Baddeley, for example, used the Society's microscope and blow-pipe to identify the minerals he had found while exploring the Lac Saint Jean area.

Baddeley also brought home for the museum some baffling fossils he'd found, which seemed to belong to "a class of animals with which naturalists are totally unacquainted." In fact they were trilobites, killed off by a mass extinction due to excess CO_2 more than 252 million years ago.

The Society's large museum occupied pride of place in the middle of Parliament House. Encyclopedic in ambition, its curators were interested in collecting everything

Blow-pipe

Chemical tests

Instructions for the management of the blowpipe, and chemical tests. John Mawe, London 1825.

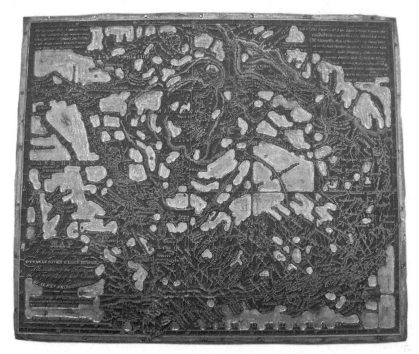

Copperplate of Alexander Shireff's maps of the land between the Ottawa River and Lake Huron, showing "land suitable for settlement." Map compiled by William Henderson for printing in the Society's *Transactions*, 1831.

under the sun. In its first few years, for example, listed donations included a section of a cow's throat, a species of intestinal worm, a white mouse, and "a singular specimen of the potato." Not surprisingly, curators complained they were having a difficult time correctly classifying and displaying all the donations.

The Society was twice ravaged by fires, and each time it had to begin its whole museum collection again. During the first fire, which burned down Parliament House, Society members ran around trying to save books while the stuffed animals, including the moose, the bear and the caribou, burst into flames. The Society library still exists after nearly 200 years, but the museum never recovered.

The be-knighted collector

James MacPherson Le Moine

After the second major fire at the Literary and Historical Society destroyed half of its library and the museum in 1862, the man who put it on its feet again was James

MacPherson Le Moine. He was in some ways the quintessential naturalist of his time—an aristocratic amateur polymath keen to advance scientific knowledge about Quebec's plants, animals and birds.

Born into a seigneurial family in 1825, Le Moine was proud to be non-denominational, bicultural, and perfectly bilingual. He moved easily between the two predominant faiths, cultures and languages of Quebec. He grew up with his English-speaking grandparents (the MacPhersons) and went to school in French at the village school. He was raised by Anglicans, baptized a Roman Catholic, had a Presbyterian wedding, a Catholic funeral, and was buried in a Protestant cemetery. He was equally at home among the French-Canadian upper classes and the British or Anglo-Quebec bourgeoisie.

As a child, Le Moine spent most of his time wandering around the woods and shores of the St. Lawrence with a neighbour, a keen birder, or fishing in the river. He developed an intense relationship with the plants and animals around him, especially the birds, and a desire to understand their lives.

He began studying law in 1846 and was later hired as a tax collector and inspector. But he saved plenty of time for pursuing his interests as a historian, naturalist and prolific writer. Le Moine's literary output was immense. His first book, *Ornithologie du Canada* (1861), was the first to popularize ornithology throughout Quebec. He went on to write 40 more books and more than 400 articles in both French and English. As well as writing about birds, he published on hunting, fishing, history, botany, French-Canadian folklore, the origins of place names, on French-Canadian customs—and swear-words—and on anything and everything else that interested him. In 1897 he was knighted by Queen Victoria for his literary contributions to Canada.

> In 1836, I was eleven years of age and though a puny, terrorized school boy, I was also the most expert squirrel, rat or bird catcher and tamer of wild animals as pets in the whole populous parish of St. Thomas.
>
> - James MacPherson Le Moine, *Souvenirs & Réminiscences*

In 1861 he wrote that Canada needed a fully equipped natural history museum, and that this would be a much better use of its money than supporting distant wars. Canada had just spent $80,000 on the Crimean War, "which the colony has no business meddling in anyway." In the meantime, he set about rebuilding the museum of the Literary and Historical Society. He joined the Society in 1863 and was an active council member for 40 years, including seven years as president and ten years as the curator of its natural history museum.

Le Moine's goal for the museum was ambitious. He said it should "procure every specimen of the Canadian fauna; second, Canadian birds and animals being complete, to obtain specimens of foreign fauna—American and European." He began to fill the museum with birds from his own collections, and bought more. He made sure it had an impressive birds' egg collection, which he considered important to the study of ornithology, and sent out calls for donations of wood, minerals and insects. He took a large selection of stuffed owls, eagles and ravens on a tour of the secondary schools of Quebec to encourage the students to visit the museum.

In addition to buying specimens, museums such as this operated on an exchange system. Just as there was a worldwide exchange of periodicals in the community of learned societies, museums and collectors offered each other duplicates of specimens in exchange for specimens they needed. In 1868, the French monk and naturalist Frère Ogérien (Jean Auguste Célestin Étienne) brought over 300 stuffed birds to exchange for Canadian birds. Le Moine was mortified that the Society had no duplicates to offer him. Together with Université Laval, the Laval Normal School (a teaching college) and other private collectors, Le Moine still only managed to scrape together enough duplicates to send the Frenchman home with half of what he'd brought. Le Moine appealed urgently to another crucial supplier of specimens, the many sportsmen who were members of the Society, to "set aside for the Society birds and animals shot by them" as duplicates available for exchange.

But the Society never had enough money or space to fulfil Le Moine's huge ambitions for the museum. And even the creatures it did have were under attack: "One or two large ones ... have been eaten by moths and insects," he reported in 1870. He longed to fill the museum with large animals, or "denizens of the forest" as he called them, but after a few years he reported glumly that he could no longer recommend buying any, as there simply wasn't enough room. He continued to advocate, without success, that the Society buy new premises to make room for gigantic stuffed animals. Now all that is left of its natural history collections is a solitary stuffed duck.

"Le cygne du Canada." Illustration by "JTW" in Le Moine's *Chasse et Pêches au Canada*, 1887.

Menura superba, or Lyre bird, by Elizabeth Gould, in *Birds of Australia*, 1840

III — DRAWING NATURE

Before photography, naturalists either were artists themselves, or needed an artist close at hand to record enough details that a certain identification could be made. Illustrations were often intentionally beautiful, and achieved an elegance and clarity that photos could not. In botanical illustrations, leaves are turned towards the observer, dead petals are removed, insect-eaten bits are repaired, and different angles can be displayed as well as every stage in the plant's life cycle.

Excluded from most other aspects of research, women were often illustrators, and have left a legacy of work that has inspired artists to be scientists, and scientists to be artists, ever since.

> The greatest flower painters have been those who have found beauty in truth; who have understood plants scientifically, but who have yet seen and described them with the eye and the hand of the artist.

> Wilfred Blunt, *The Art of Botanical Illustration*, 1950

Drawing dissected mollusc penises

Natural history artists

People have been depicting the natural world for millennia. Some of the earliest known scientific illustrations were of medicinal plants. Books called herbals described the appearance and properties of plants and gave practical advice on how to prepare medicines from them.

One Greek herbal, written in 65 BC, was translated into Latin and Arabic and was the most influential work on medicinal plants in the Christian and Islamic worlds until the late seventeenth century. In 1552, Martinus de la Cruz, an Aztec scholar in Spanish-ruled Mexico, created an Aztec herbal illustrating plants that were said to cure, among other ailments, black blood, difficulties with urine, weakness of the hands, stupidity of the mind, and "goaty armpits of sick people."

Later, natural history artists illustrated books about anatomy, cooking, horticulture, and mapmaking.

The explorers of the eighteenth and nineteenth centuries, along with Carl Linnaeus' classification system, changed the way artists represented the natural world. Before

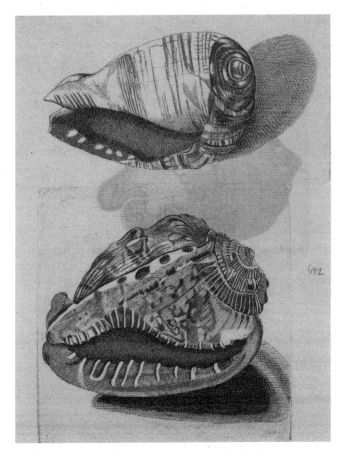

Engravings of shells by Anna and Susanna Lister, in Martin Lister's *Historiae Conchyliorum*, 1685.

this, artists new to the Americas drew the creatures and plants they found based on memory, on an ideal, or on what was already known. But once Linnaeus began categorizing plants according to their visible parts, creating a single hierarchy, artists gradually began to record precisely what a flower (and no other flower yet known) looked like so it could be identified with certainty. This new belief in "the visible as the ground of truth" became the credo of the eighteenth century. Linnaean botany mostly ignored other aspects of plants, however, such as their location, local uses or local names.

Long before photography or refrigeration, explorers hired artists to accompany their voyages to make an on-the-spot record of specimens before they rotted, were gobbled up by maggots, or were lost through disaster. So they had to be quick. Sydney Parkinson, one of Joseph Banks' artists on the *Endeavour*, made 1350 drawings in two years and four months (1768-1771). He worked in a frenzied rush, sitting up all night in a tiny cabin surrounded by specimens. He died before the voyage was over. Back home in Britain, Banks employed 18 engravers to produce 753 copper plates from Parkinson's drawings.

The work of cartographers or natural history artists often combined Indigenous with European knowledge. On the same voyage, Tupaia, a Tahitian navigator, worked with James Cook making maps of 74 islands. Many of Cook's notes about depth, rocks, landmarks and shelter from winds were likely informed by Tupaia, who navigated the ship for weeks. Tupaia recited inherited lists of landmarks from memory, as did all Polynesian navigators. When he died of a fever, Cook wrote in his journal "He was a shrewd, sensible, ingenious man, but very proud and obstinate."

Red Poppy by Elizabeth Blackwell, in *A Curious Herbal*, 1735

The work of scientific artists, especially when they were women, was often in the shadows. The seventeenth-century British doctor and naturalist Martin Lister, unsatisfied with the work of contemporary artists, trained his young daughters Susanna and Anna to observe, draw, etch, engrave, use a microscope, and possibly to dissect: Susanna recorded in her notebook that she had drawn "a brachiopod gill and dissected mollusc penises." Their father's *Historiae Conchyliorum* (1685), a compendium of all known shells, was printed on a press at home by his son and is filled with hundreds of beautiful illustrations by his daughters.

For some women, botanical illustration was a way to make a living, or in Elizabeth Blackwell's case, to post bail. In 1739, she published a herbal of 500 drawings to rescue her husband from debtor's prison. She engraved the drawings on copper, printed them, painted them and then took them to the prison, where her husband added the plant names in Latin, Greek, Italian, Spanish, Dutch and German.

Elizabeth Coxen (1804–1841) was finding her job in London as a governess "wretchedly dull" until she met her future husband, John Gould, in an aviary, and began illustrating his ornithology books. Her bird illustrations gained her international fame, and she illustrated the birds collected by Darwin on his *Beagle* expedition. In 1838, she and the eldest of their six children accompanied Gould to Australia, where she illustrated his books on birds and kangaroos. She died aged 37 giving birth to her eighth child. Although her husband named a finch, *Erythrura gouldiae*, in her memory, her name does not appear on his books. He never found another artist her equal.

Contrary to the Linnaean mode of isolating plants from their environment, Marianne North (1830–1890) insisted on painting plants in their natural settings. When her parents died, she sold the family home and began travelling to paint the flora of various countries. Between 1871 and 1885 she visited the US, Canada, Jamaica, Brazil, Tenerife, Japan, Singapore, Sarawak, Java, Sri Lanka, India, Australia, New Zealand, South Africa, the Seychelles and Chile. She travelled alone, much to the disapproval of her male friends.

In the nineteenth century, natural history artists began to aspire to objectivity. Wary of human mediation between nature and representation, by the mid-nineteenth century scientists began to think that mechanically produced images such as photography were more trustworthy.

While her husband mapped the river

Natural history artists in the Canadas

Inuit had been carving masks and figurines for spiritual or religious purposes out of walrus tusk or bone since time immemorial. In the early nineteenth century, they traded dolls, toys and animal carvings with whalers, sailors and explorers. After 1948, Canada began to encourage carving as a source of income, and Inuit started to sculpt out of serpentinite and soapstone. Their vivid and accurate portrayals of birds, bears, walruses and seals were the fruit of the many hours of patient observation required of the best hunters. They also portrayed mystical creatures which to them were as real as the other animals and birds.

Few of the first Europeans who illustrated the creatures of Canada had actually been there, so they drew specimens sent back by explorers, improvised from textual accounts, or else shamelessly copied from others. Herman Moll's picture of beavers on his 1720 map, for example, was copied from a 1698 chart by Nicolas de Fer, who had copied his beavers from books by Louis Hennepin (1697) and François Du Creux (1664). Herman Moll's copied map, in turn, was pirated by Jonathan Swift for *Gulliver's Travels* in 1726. One exception to all this copying was Jesuit missionary Louis Nicolas, whose 180 enchanting drawings in *Codex canadensis* (1700) were mostly of plants and animals he had seen, although he includes a unicorn and a merman. Pre-Linnaeus, Nicolas showed only the edible or useful parts of plants and didn't bother with the rest.

Anything that lived in water was considered a fish, and animals were mostly depicted as if dead, their bodies stretched out and stiff. This is because at the time, animals were most useful to science when they were dead. It would take a radical change in perspective for animals to be drawn in a more realistic way. It was only when people began to question why a certain characteristic was useful to the animal itself, rather than to humans, that the old anthropocentrism would shift.

New France did not produce its own engraved imagery, as it lacked both the market and the presses required. With British settlement came printers and publishers, artists and engravers, and a larger reading public.

Marianne North in Sri Lanka, 1877.

An industrious beaver colony near Niagara Falls. Inset from *The World Described; or a New and Correct Sett of Mapps Shewing the Several Empires, Kingdoms, Republics, Principalities, Provinces &c in all Known Parts of the World.* Herman Moll, 1715.

Many of the wives of British officials in Quebec had studied botany, gardening, and illustration as part of their general education, and were keen to find and draw Canadian plants. Christian Dalhousie, Harriet Sheppard and Anne Perceval drew pictures of the plants they found while "botanizing" together on their large estates (see page 35). Fanny Amelia Bayfield (1813-1891), while her husband Henry Bayfield surveyed the Saint Lawrence River and the Gulf, painted exquisite watercolours of minutely-observed plants, butterflies and insects. She bequeathed them in an album, *Canadian Wildflowers*, to her son on her death. Another talented botanical illustrator was Ann Ross McCord (1807-1870), matriarch of the wealthy McCord family in Montreal.

These women were reticent about taking credit for their work. Amelia Bayfield did not even sign her paintings. Botanical illustration offered professional possibilities that were usually reserved for men, but women were encouraged to teach botany or to illustrate, rather than to participate in higher scientific circles. William Dawson, Canadian palaeontologist who published hundreds of scientific works, (see page 64), made a point of teaching his daughter Anna to "sketch from nature" and his books are full of her drawings and wood engravings. The first botany degree to be awarded to a woman was in 1894.

For other women, botanical illustration was a matter of survival. Agnes Fitzgibbon, daughter of Susanna Moodie (whose *Roughing It In the Bush* is an account of early settlement in Upper Canada) illustrated and published a botanical book by her aunt, Catharine Parr Traill, to support her six children after her husband died. She paid

From Louis Nicolas' drawings of Canadian fauna, *Codex canadensis*, ca. 1700.

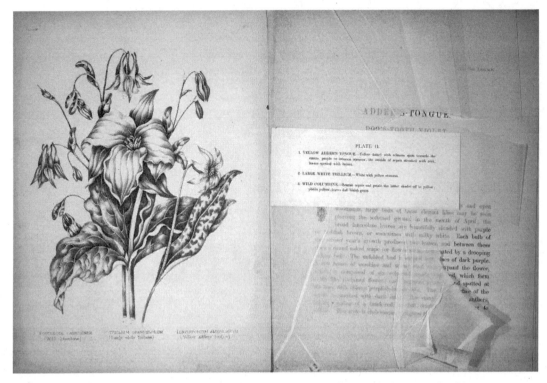

Wild columbine, large white trillium and yellow adder's tongue, with colour notes, by Agnes Fitzgibbon. In *Canadian Wild Flowers* by Catharine Parr Traill. Montreal, Quebec, 1869.

her aunt $50 for the text. Then she enlisted a printer, launched a campaign to attract 500 subscribers (an early version of crowdfunding), and drew all the flowers. She had borrowed a block of limestone from a printer, on which she taught herself lithography in order to print the 500 plates of her drawings, which she then coloured by hand. The book, *Canadian Wild Flowers*, sold out within six weeks and went through four editions in as many decades.

Others again were drawn to natural history illustration through hunting, fishing and farming. In 1897, Johan Beetz (1874-1949), a Belgian aristocrat who had hunted and fished his way around Africa, built a 12-room wooden house at Piastrebaie on the North Shore of the Saint Lawrence, 60 km east of Havre Saint-Pierre. He spent 50 years there, hunting, trapping and fishing with the local villagers, one of whom, Adéla Tanguay, he married. He was a fox farmer, and his doctoral thesis on the topic included 1800 anatomical drawings of foxes. Later he moved to Montreal and then to the city of Quebec, where he co-founded the Quebec Zoo in Charlesbourg on one of his fox farms. The zoo closed in 2006 after 75 years of operation.

IV – EXPLAINING NATURE

Whether it was Haudenosaunee leader Domagaya giving Jacques Cartier the medicine for scurvy, French doctor Michel Sarrazin sending plant specimens from Quebec to the Jardin des Plantes in Paris or Scottish-educated William Dawson ruminating on the fossils in Nova Scotia, the connections between Europe and the Americas generated new knowledge about our global ecological system, or the biosphere, as we sometimes call the natural world.

Naturalists were prepared to suffer to make these connections—violent seasickness (Charles Darwin), shipwrecks (Alfred Russel Wallace), or even death (illustrator Sydney Parkinson). What drove these nineteenth-century naturalist-explorers to spend so many miserable and sometimes dangerous weeks crossing oceans? "The misery I endured from sea-sickness is far far beyond what I ever guessed at," wrote Charles Darwin to his father during the voyage of the *Beagle*.

Travelling to collect specimens was a rite of passage for naturalists—a form of pilgrimage. Naturalists and artists rode on the coat-tails of imperial expansion and colonial

P.Z.S. 1904, vol. II. Pl. XXIV.

H. Grönvold del. et lith.

Mintern Bros. imp.

SIMIA VELLEROSUS (Gray)
(very old male)

Simia vellerosus (Gray), Northern Cameroon. Drawn and lithographed by H. Gronvold, 1862.

commerce, joining the exploration parties, compulsively collecting anything they could find and sending sketches and specimens home. Meanwhile, Indigenous people, the original inhabitants of the Americas, paid a much higher price for these explorations, dying of the diseases the newcomers brought or getting involved in their deadly wars.

Geologists analyzing different layers of Canadian rock and their respective fossils wondered, as did their European counterparts, why species seemed not to be static, exactly as God had created them, but to have changed over geological time. Biologists wondered why species were so different in different places.

Various theories were proposed, but in 1858, two travelling naturalists suggested a plausible reason. One a parson and the other an animal trader, they independently came to the same worldview-shattering conclusion after sailing across the ocean to investigate species variations amongst isolated islands. Their theory provoked violent opposition on both sides of the Atlantic, and changed the practice of science for ever.

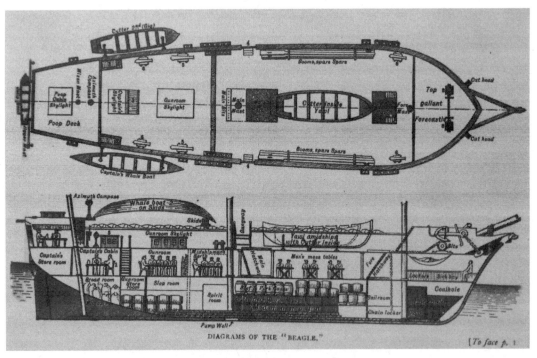

HMS *Beagle* based on a drawing from memory by Philip Gidley King, midshipman on the second voyage, 1832.

Com̃ment noel après le deluge annua a terre et mist hors le bestail et fist sacrifice et planta la vigne⁖

1. **Exit from Noah's Ark. Paris, Bedford Group, ca. 1420**. Discoveries of fossils of unknown animals in successive layers of rock made naturalists wonder what had caused these extinctions. They tried to harmonize their discoveries with the Biblical account of the Flood and Noah's Ark. Did Noah forget to save a few creatures? In this painting, the ravens are eating the drowned sinners and Noah is already drunk.

2. **Helianthea osculans (buff-tailed star-frontlet hummingbirds), by William Matthew Hart, 1880. Illustration in John Gould's *A monograph of the Trochilidae, or family of humming-birds.*** John Gould mounted a display of 1500 stuffed hummingbirds at the Great Exhibition in London in 1851. Found only in the Americas, they quickly became a European fashion craze. In 1888 an auction house in London sold 400,000 hummingbird skins in a single afternoon. The movement to halt the wearing of feathers was the origin of the Royal Society for the Protection of Birds, now the largest nature conservation charity in Europe.

Plate 473.

3. **Elecampane (*iluna helenium*, or horse-heal), by Elizabeth Blackwell in *A Curious Herbal*, 1739.**
Elizabeth Blackwell drew, engraved and hand-coloured this and hundreds of other medicinal plants from life and published the herbal to get her husband out of jail. Her husband was later hanged for treason in Sweden. Elecampane has been highly valued since Roman times: Pliny wrote that it helped digestion, caused mirth, and that the root "doth fasten the teeth." Modern research has shown it has potential for controlling staphylococcus infections.

The Cataract of NIAGARA, some make this Water-fall to be half a League while others reckon it no more than a hundred Fathom.

A View of ỹ Industry of ỹ Beavers of Canada in making Dams to stop ỹ Course of a Rivulet, in order to form a great Lake, about w.ᶜʰ they build their Habitations. To Effect this; they fell large Trees with their Teeth, in such a manner, as to make them come Cross ỹ Rivu-let, to lay ỹ foundation of ỹ Dam: they make Mortar, work up, and finish ỹ whole with great order and wonderfull Dexterity. The Beavers have two Doors to their Lodges, one to the Water and the other to the Land side. According to ỹ French Accounts.

4. **Inset from *The World Described; or a New and Correct Sett of Mapps Shewing the Several Empires, Kingdoms, Republics, Principalities, Provinces &c in all Known Parts of the World*, Herman Moll, 1715.** As well as being economically important, beavers had a reputation as the master craftsmen of the animal kingdom. Note that the lion-faced beavers are walking upright, organized into strict work teams, and use their tails to carry the mortar. Herman Moll never been to Niagara Falls or seen a beaver.

5. ***Menura superba*, or lyrebird, by Elizabeth Gould, in *Birds of Australia*, 1840.** John Gould's description of Australia's shyest bird focuses on how difficult it is to shoot. He had to stick male lyrebird feathers in his hat and waggle them while crouching in thick bushes. With a well-trained dog, though, it was "easily approached and shot," or with the help of Aboriginals with their "noiseless and gliding steps." Lyrebirds can mimic a variety of birds and dogs as well as artificial sounds such as cameras and chainsaws.

P.Z.S.1904, vol. II. Pl. XXIV.

H. Grönvold. del. et lith.   Mintern Bros. imp.

SIMIA VELLEROSUS (Gray)
(very old male)

6. **Simia vellerosus (Gray), Northern Cameroon. Drawn and lithographed by H. Gronvold, 1862.** Many people, whatever their religious beliefs, welcomed the theory of evolution by natural selection—except for the part about us humans being so closely related to apes. Geologist William Dawson, Principal of McGill, objected to "this endless pedigree of bestial ancestors, without one gleam of high or holy tradition to enliven the procession."

7. **"HMS Erebus in the Ice, 1846," by François Étienne Musin.** The aim of Franklin's expedition was to traverse the last unnavigated sections of the Northwest Passage in the Canadian Arctic and to record magnetic data to help navigation. The ships became icebound in Nunavut and not a single member of the expedition survived. Many expeditions were sent to find the wreck. In 1869, Inukpujijuk, an Inuk, drew a map showing its location; but it wasn't until 2014 that the wreck was discovered, very close by, in a strait called Umiaqtalik, which means "There is a boat there." In nearly 200 years no one had thought to ask.

8. **Officer's Room in Montreal, by Cornelius Kreighoff, 1846.** British officer, surgeon, artist, writer and naturalist Andrew Aylmer Staunton created his own private museum or *cabinet de curiosités*—stuffed birds, Indigenous-made objects, sports equipment such as bridles and guns, and paintings (some by Krieghoff)—which he had collected during his stay in Canada.

Seasick on the *Beagle*

Origins of the *Origin…*

As a young man, Darwin was quietly studying at Cambridge to be a parson, a job considered a safety net for second sons to prevent them from squandering their father's fortunes. But he was also an amateur naturalist, as were many English parsons in the early nineteenth century. He spent hours collecting beetles and studying geology, and he longed to see more of the world before settling into life in a parsonage. Through his connection with the Cambridge professor of botany, John Stevens Henslow, he was offered the chance to travel as a naturalist and companion to Robert FitzRoy, captain of HMS *Beagle*. Fitzroy wanted a companion to break the isolation, one who shared his scientific interests but who was also a "gentleman" with whom he could dine as an equal.

In 1831, at the age of 22, Darwin left on a five-year voyage around the world.

Darwin had prepared himself for loneliness. "The mind requires a little case-hardening," he wrote back to Henslow, "before it can calmly look at such an interval of separation from all friends." Yet he later wrote that "The voyage of the *Beagle* has been by far the most important event in my life and has determined my whole career."

It was a tiny British Navy ship, only 90 feet long, and made of oak—very possibly oak from the St. Lawrence Valley, as the Navy had relied on Canadian timber supplies since Napoleon cut off its supply from the Balkans. Since the end of the Napoleonic war, Britain had been using its warships to chart the oceans and coasts of the world. In the interests of empire and free trade, the *Beagle*'s mission was to map and make hydrographic and meteorological surveys of the southern coasts of South America. A new evolutionary theory was its unintended consequence.

Darwin was seasick for the entire time at sea, and living in very cramped quarters. He only really felt comfortable when lying in his hammock, but his face was only two feet from the deck above. "Oh a ship is a true pandemonium," he wrote in his diary, "& the cawkers who are hammering away above my head veritable devils." When he arrived home he had lost over six kilos.

On land, however, he never rested. In a single day he might catch 68 species of beetle, shoot 80 species of birds and come back aboard the *Beagle* carrying a fossilized giant ground sloth bought from a gaucho (South American cowboy) for two shillings. Then he

In the Bay of Biscay there was a long & continued swell & the misery I endured from sea-sickness is far far beyond what I ever guessed at….Nobody who has only been to sea for 24 hours has a right to say, that sea-sickness is even uncomfortable.— The real misery only begins when you are so exhausted—that a little exertion makes a feeling of faintness come on.— I found nothing but lying in my hammock did me any good.— I must especially except your receipt of raisins, which is the only food that the stomach will bear.

- Letter to his father, February 8th 1832.

got busy describing, packing, numbering and shipping the skins, bottles and bones back to Cambridge. He had help from Syms Covington, "fiddler and boy to the poop cabin," who was his assistant, secretary and servant until 1839, when Covington migrated to Australia. Darwin found time to write a 189,000-word journal, which he sent home in instalments to be read aloud by the family.

Darwin didn't deny some of his weaknesses. He worried about his enthusiasm for slaughtering animals, his minimal language skills, and his smoking. But he impressed the Uruguayans no end by striking matches with his teeth.

HMS *Beagle* returned to England via Tahiti and Australia after circumnavigating the world. His friend Adam Sedgwick wrote of Darwin, "There was some risk of his turning out an idle man: but his character will now be fixed, & if God spare his life, he will have a great name among the naturalists of Europe."

Scientists are often said to make the same discoveries and inventions independently and almost simultaneously in different parts of the world: Galileo and Scheiner hit on sunspots, Leibniz and Newton hit on calculus. Darwin was not the only one who hit on evolution by natural selection.

Feel it struggling between one's fingers

Alfred Russel Wallace and the theories of evolution

Darwin developed his theory of evolution by natural selection over many years back in the comfort of his house in Kent, England, long after his voyage on HMS *Beagle*. Alfred Russel Wallace had almost the same idea, but it came to him in the middle of his own voyage of exploration while he was lying on a bamboo bed with a tropical fever on an island in Indonesia. Although there is only one *Origin*, there are two discoverers of the theory, one of whom is scandalously under-celebrated.

Recalling his eureka moment years later, Wallace wrote "Why do some die and some live? And the answer was clearly, that on the whole the best fitted live ... Then it suddenly flashed upon me that this self-acting process would necessarily improve the race, because in every generation the inferior would inevitably be killed off and the superior would remain—that is, the fittest would survive."

In 1858 he sent his ideas in essay form to Darwin, who had not yet published his own. Darwin was astonished. "I never saw a more striking coincidence. If Wallace had my manuscript ... he could not have made a better short abstract! Even his terms now stand as heads of my chapters." Wallace's essay was published, along with a description of Darwin's own theory, in the same year. Against all odds, until the end of their lives there was little bitterness or rivalry between them.

Flying frog, by J. G. Keulemans, from a drawing by A. R. Wallace. In *The Malay Archipelago: The land of the orang-utan, and the bird of paradise. A narrative of travel, with sketches of man and nature,* 1869

Unlike Darwin, who was independently wealthy, Wallace left school at 13 and educated himself in Mechanics Institutes, working-class adult education institutions usually with a library and a museum. While working as a builder, watchmaker, and land surveyor, he became a keen naturalist, reading geology books by Charles Lyell and on Lamarck's theory of the "transmutation" of species, or how one species changes into another over time.

While Darwin called himself "a machine for generating hypotheses," Wallace was a taxonomist and a species hunter. With the Victorian mania for collecting fossils, animals, insects, shells and plants, there was money to be made collecting them in remote parts of the world and selling them to museums and wealthy amateurs wanting to add to their private museums or "cabinets." Inspired by Ida Laura Pfeiffer (1797–1858), who made a solo trip around the world and financed herself by selling specimens and writing books, Wallace set off for Brazil at the age of 25 with his friend Henry Bates. For four years, he made careful note of his observations on transmutation of species, and collected specimens worth more money than he'd ever made before.

On his way home, the ship caught fire. Wallace only just escaped with his life, and spent ten days in a lifeboat. All his specimens were lost, including live birds and monkeys he had been taking care of for years.

I trembled with excitement as I saw [a butterfly] come majestically toward me & could hardly believe I had really obtained it till I had taken it out of my net & gazed upon its gorgeous wings…. I had seen similar insects in cabinets, at home, but it is quite another thing to capture such one's self – to feel it struggling between one's fingers, and to gaze upon its fresh and living beauty, a bright gem shining out amid the silent gloom of a dark and tangled forest.

- Wallace, A.R. (1869). *The Malay Archipelago: The Land of the Orang-utan, and the Bird of Paradise: A Narrative of Travel, with Studies of Man and Nature.*

A perfect stranger to discouragement, Wallace set off again two years later, this time in a crowded steamship. There were more animals aboard than humans. Passengers had to bring their own deckchairs, and slept four to a berth. Wallace played chess all the way. Then he spent eight years collecting and exploring in the Malay Peninsula, or what is now Indonesia, Malaysia and Singapore. After 14,000 miles of travel, Wallace came home with 126,000 biological specimens, including a thousand species new to science.

Wallace is the founder of biogeography, or the study of the geographical distribution of living things. He noticed that the animals and plants on the islands were split between Asian and Australian types, even though they might be only separated by a few miles. This divide, where the continents once split apart and drifted away from each other, is known as the Wallace Line. The sharp differences in the animals on each side confirmed his emerging theory of evolution by natural selection.

Wallace had many other interests. He saw clearly that human activity was damaging the environment and would result in extinction of species and widespread pollution. He was a pacifist and a socialist, and believed that land was a public resource that should be nationalized. He also believed in the vote for women and, unlike many of his contemporaries, believed that the minds of non-Europeans were every bit as capable as those of Europeans. "The more I see of uncivilized people," he wrote, "the better I think of human nature on the whole, and the essential differences between civilized and savage man seem to disappear." He was also vehemently opposed to the newly-introduced smallpox vaccination laws. His belief in spiritualism, however, including in communication between the living and the dead, lowered his credibility in the scientific community.

> To pollute a spring or a river, to exterminate a bird or beast, should be treated as moral offences... never before has there been such widespread ravage of the earth's surface by destruction of native vegetation and with it of much animal life, and such wholesale defacement of the earth by mineral workings and by pouring into our streams and rivers the refuse of manufactories and of cities; and this has been done by all the greatest nations claiming the first place for civilization and religion!
>
> - Wallace, A.R. (1910). *The World of Life.*

Wallace was prodigious in his discoveries. Over 300 species bear his name. He was also a prolific writer, publishing 645 articles and 20 books, while Darwin published only ten books. His writings transport the reader into the sensuality and challenges of specimen hunting. The spiders in the Malay jungles, he wrote,

> ....were a great annoyance, stretching their nets across the footpaths just about the height of my face; and the threads composing these are so strong and glutinous as to require much trouble to free oneself from them. Then their inhabitants, great yellow-spotted

monsters with bodies two inches long ... are not pleasant things to run one's nose against while pursuing some gorgeous butterfly, or gazing aloft in search of some strange-voiced bird.

"I laughed ... till my sides were almost sore"

How the theory of evolution by natural selection was received

The theory proposed by Wallace and Darwin was a shock to people's understanding of where human beings come from, which was mostly based on the Biblical account in Genesis. However carefully they presented their evidence, then as now, not everyone was prepared to accept it.

Perhaps knowing it would be controversial, or perhaps from inherited lethargy, Darwin waited 20 years before publishing his *Origin of Species* in 1859. But a Scottish journalist had hit on a similar theory and beat him to it.

In 1844, Robert Chambers anonymously published *Vestiges*, a wildly popular book (Queen Victoria read it out loud to the Prince Consort) suggesting that humans had evolved from apes. Darwin's friend the geologist Adam Sedgwick called it a "foul book" in which "gross credulity and rank infidelity joined in unlawful marriage." Novelist George Eliot, however, who was also an amateur biologist, admired it. *Vestiges* prepared the ground for the shock of Darwin's much better argued book.

> I have read your book with more pain than pleasure. Parts of it I admired greatly; parts I laughed at till my sides were almost sore; other parts I read with absolute sorrow; because I think them utterly false & grievously mischievous.
>
> - Letter to Darwin from geologist Rev. Adam Sedgwick

Some British scientists, such as Thomas Huxley, welcomed Darwin's theory for the very reason that it undermined literal belief in the Old Testament and challenged the moral authority of the clergy. Others found that it strengthened their faith: Charles Kingsley wrote that "God's greatness, goodness and perpetual care I never understood as I have since I became a convert to Mr Darwin's views."

While many German and French naturalists attacked Darwinism, some Christian and Muslim intellectuals in the Arab world embraced it. It was a challenge to repressive Ottoman imperial traditions, and was conscripted in support of pan-Arab secularism, giving Arabs a way to broaden their interpretation of scripture and theologies about the physical universe. Other Arab religious thinkers, both Muslim and Christian, attacked Darwin on purely religious grounds.

> Mr. Darwin is most likely right in his opinion, but I doubt it.
>
> - John Macoun, Canadian biologist, letter to Joseph Hooker, Director of Kew Gardens, 1866.

Caricature of Darwin as a monkey in a Parisian satirical magazine, André Gill, 1878-9

The opposition to Darwin in Canada, too, was principally religious. It appeared to contradict the belief that God created human beings. The University of Toronto rejected two of Britain's most eminent scientists—Thomas Huxley and John Tyndall—as potential faculty because of their "extreme" evolutionist positions. Two of Canada's own most important scientists spent a lifetime arguing against the theory, and both were in Quebec: geologist William Dawson and priest and naturalist Léon Provancher.

A chaos of fallen rocks

Passionate opposition in Quebec

Some of the most virulent opposition in Canada came from Catholic and Protestant Quebec.

Shortly after *Origin of Species* was published, Pope Pius IX riposted with *Syllabus of Errors* (1864), a wide-ranging attack on modernity. This gave the Ultramontane Catholic authorities in Quebec the excuse to shut down any discussion of evolution, and they excommunicated members of a liberal learned society in Montreal, the Institut Canadien.

Geologists at Université Laval, Abbés Thomas-Étienne Hamel and Joseph-Clovis-Kemner Laflamme, had to tread carefully. Laflamme lectured in 1877 that Darwinism was "an absurdity," while Hamel presented the theory in his lectures without overtly taking a side. They both argued on scientific grounds that the theory did not follow the correct Baconian scientific method of inducing theory from accumulated observations, as opposed to testing a hypothesis. As time went on, the theory of evolution by natural selection became more difficult to argue with, and Laflamme in 1907 conceded that there was doubtless a *grande loi de perfectionnement* ("great law of development"), but still not applicable to people, who came out ready-made and with a soul.

Abbé Louis-Ovide Brunet, who taught botany at Université Laval, was no fan of evolution either, but shared plant specimens with colleagues at Harvard. They wanted them because Canada's plant distribution patterns were important to a better understanding of Darwinian theory. Had Brunet known what his specimens were being used for, he might have thought twice. Meanwhile his friend the entomologist William Couper, whose collection of 6000 insect specimens were on display at the natural history museum of the Literary and Historical Society of Quebec, openly supported Darwin's theory and lectured on it enthusiastically in Montreal in 1875.

Catholic priest and naturalist Léon Provancher (1820-1892) began taking an interest in science as a child when he saw a shellfish fossil discovered by workmen digging a well. In his parishes of Grosse Ile, Isle-Verte, and Saint-Joachim, he began to study the

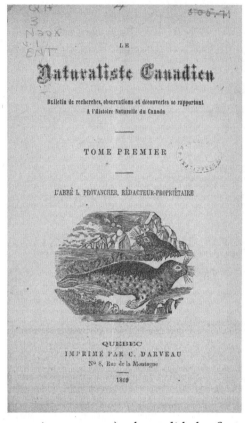

First issue of *Le Naturaliste canadien*, 1868

flora and fauna along the St. Lawrence River. In 1862 he published his two-volume *Flore canadienne*. The same year, as an expert in horticulture, he published a book on how to grow fruit and vegetables in Lower Canada. He was also an entomologist, and discovered more than 1000 species of sawflies, wasps, bees, and ants, naming many of them after his friends. Provancher's wide-ranging interests also included a passion for scientific education. In 1868 he founded a *Le Naturaliste canadien*, a popular science journal that is still publishing to this day.

Provancher was fiercely opposed to Darwin's theory of natural selection, and dedicated 12 articles in his scientific journal to attacking the theory and ridiculing evolutionists. Most of his attacks were on the grounds of "impiety," but his more scientific questions were, a) where did the first organism come from? b) if our own efforts at deliberate selection (e.g. dog breeding) fail to change one species into another, how can natural selection, which is more haphazard, succeed? c) the mummies of Egypt are 3000 years older than us, so why are they recognizably human? By popular demand, Provancher was still giving courses at Université Laval arguing against Darwinism as late as 1887.

An equally devout opponent was the Presbyterian geologist and pioneering paleobotanist John William Dawson (1820-1899), who had discovered the earliest reptilian remains then known in North America. He published more than 400 books and articles, and was later principal of McGill (see page 94). He sent specimens to Darwin, who co-signed a testimonial for Dawson's election to the Royal Society of London. In 1858 he found a supposed fossil, *Eozoön canadense*,

The evolutionist, instead of regarding the world as a work of consummate plan, skill and adjustment, approaches nature as he would a chaos of fallen rocks, which may present forms of castle and grotesque profiles of men and animals, but they are all fortuitous and without significance.

- William Dawson, in *The Story of Earth and Man*, 1873

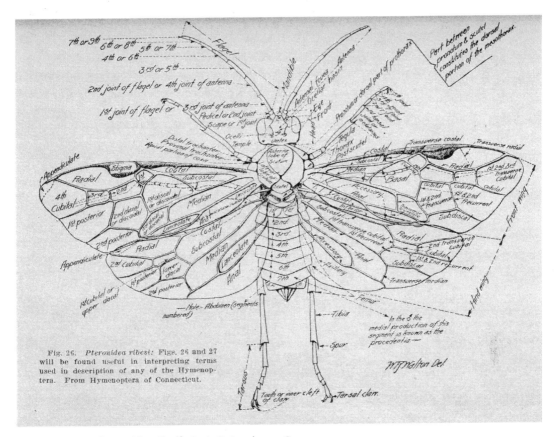

Sawfly wing parts (*Pteronidea ribesii*) , by A. G. Ruggles, 1918.

which he believed was evidence of the existence of a relatively sophisticated organism that predated simpler forms—and therefore disproved evolution by natural selection. He published with London's Geological Society on *Eozoön* as Canada's first life-form. Most other scientists, however, correctly believed it was non-organic, including the official palaeontologist at the Geological Survey of Canada (see page 89), Elkanah Billings.

Like Provancher, Dawson thought Darwinism "reduced nature to an automaton and removed the benign influence of the Supreme Being." He believed it would destroy religious beliefs and social morality. He also thought that it was Darwin's godless view of nature that was leading to degradation of the environment, a view whereby humanity is seen as "an enemy of wild nature, so that in those districts which he has most fully subdued, many animals and plants have been exterminated."

Provancher and Dawson may have avoided a pernicious misinterpretation of the theory of evolution that persists to this day, a version saying that evolution as a story of adaptation was also a story of progressive refinement, and that some humans (especially White upper-class Europeans), were more advanced than others. This was the view of Thomas Huxley, who had embraced Darwin's theory with such enthusiasm. Darwin's theory in fact implies that evolution occurs without purpose or advancement: species produce random mutations, some of

> In less than a quarter century, this absurd theory will have played itself out and its only adherents will be a few corrupt souls for whom to be counted among the brutes does not shame their appetites or aspirations.
>
> - Léon Provancher, *Le Naturaliste canadien*, 1887

which turn out to be beneficial for survival in a given environment, where descendants with the beneficial mutation gradually replace those without it.

Franz Boas (1858-1942), an American anthropologist, realized the error after Inuit saved his life in an Arctic storm. Far from being "primitive," Inuit had spent thousands of years adapting perfectly to an exceptionally harsh landscape. Genome research has confirmed that race may be a cultural or political construct, but it is not a biological one. Almost all human DNA, or 99.9 percent of the three billion base pairs, does not vary from person to person. But the revolutionary concept that we are cut from the same genetic cloth, in spite of our cultural differences, has yet to penetrate our thick skulls. Learning from naturalists in cultures that Western society has viewed as "primitive," many of which are fast disappearing, may help save all our lives in the coming storm.

V – MAPPING NATURE BY BOAT

While many of the natural history observations in nineteenth-century Canada were made by trudging across the continent on foot or shooting rapids in a hand-made canoe, others were made from the deck of a ship. This section tells three lesser-known stories of maritime explorers in Canada. None of them had much time for the schoolroom. James Cook left school at 12, Joseph-Elzéar Bernier started out his career as a cabin boy, and Robert Bartlett dropped out of college to go to sea. Indigenous knowledge was essential to the careers of all three men. Cook was later guided through the Pacific by an Indigenous Polynesian navigator. Both Bartlett and Bernier were aided, advised, supplied and guided by Inuit.

"HMS *Erebus* in the Ice, 1846," by François Étienne Musin.

No room for idlers

Captain Cook in Quebec

Captain James Cook is famous for his explorations on HMS *Endeavour*, which circum-navigated the globe visiting South America, Tahiti, New Zealand, Australia and the Indonesian island of Java. From this voyage he and naturalist Joseph Banks brought back thousands of specimens and drawings of plants and animals previously unknown to Europeans, and the expedition became a model for later scientific explorations, including Darwin's voyage on the *Beagle*.

A little-known episode of Cook's career was how he learned to make maps on his way to fight the Battle of Quebec in 1759.

James Cook was born in 1728 in Yorkshire, in a thatched cottage made of clay. His father was a Scottish immigrant farm labourer. He left school at 12, and at 17 apprenticed in the merchant navy, ferrying coal up and down English coasts in a boat called *Freelove*.

In 1755 he joined the Royal Navy. Britain was embroiled in the global Seven Years War, including French-English fighting over the colonies in North America. Having risen up quickly through the ranks, Cook set off as master (in charge of navigating and handling the ship) of the 60-gun *Pembroke* to join the war. London wanted to capture the two great French fortresses of Louisbourg, in Nova Scotia, and Quebec.

Whenever I could get a moment of time from my duty, I was on board the *Pembroke* where the great cabin, dedicated to scientific purposes and mostly taken up with a drawing table, furnished no room for idlers. Under Capt. Simcoe's eye, Mr. Cook and myself compiled materials for a Chart of the Gulf and River St. Lawrence… with no other alterations than what Mr. Cook and I made coming up the River.

 - Samuel Holland, letter to Lt.-Gov. John Graves Simcoe about the winter of 1758 in Halifax

On the way, 26 of his crew died of scurvy. Although James Lind had discovered that lemons cured scurvy in 1747, it was another 50 years before the Royal Navy issued lemon juice to its seamen. The crippled *Pembroke* crew didn't participate in the battle but looked on as Louisbourg was captured. A few days later, the military engineer Samuel Holland, who was walking about mapping the area with a surveying instrument called a plane table, noticed Cook following and watching him. Cook asked Holland to teach him to use it. Holland found him a quick study. Their next destination was Quebec, but there were no reliable charts of the St. Lawrence. And so the two men set about trying to make some.

Cook's ship was then sent on Jeffrey Amherst's expedition to destroy French-Canadian villages up and down the St. Lawrence Estuary. Whether Cook joined in burning farms and slaughtering livestock is unknown. But with his newly learned surveying skills, he charted Gaspé Bay—the first of his known charts.

Back in Louisbourg, he and Holland spent the winter on board the *Pembroke* trying to chart the rest of the river, mostly using captured French charts. In the spring James Wolfe arrived, and they set out for Quebec, correcting their charts as they went along. The hardest part of the river was *La Traverse*, between Île aux Coudres and Île d'Orléans, where the only navigable channel is very narrow with rock ledges on each side, large tides and fast tidal currents. The French were sure they were safe, believing that the big British ships would never make it through. But sounding the waters with lead plumblines as they went, and guided by captured pilots (perhaps trained at the Séminaire de Québec), Cook marked the passage with buoys and small boats to signal the way.

More than 100 ships made it safely through and anchored at Quebec. Cook made a mortal mistake when he advised Wolfe that they could easily land in flat-bottomed boats ("Whitby cats") at Beauport. They tried it, but ran aground and many men were killed. That's when Wolfe decided to attack from further north, a tactic that won him the war.

Cook's "New Chart of the River St. Lawrence" was published in London in 1760 and was used by generations of seafarers afterwards. He spent the next five years mapping Newfoundland. In the summers he went home to England, where he met his wife Elizabeth Batts, the daughter of an innkeeper in Wapping. They had six children, none of whom survived long enough to give them grandchildren.

Cook returned to England for the winter of 1767–68, his Newfoundland charts still unfinished. But his work had set a new standard for British hydrographic surveys, and

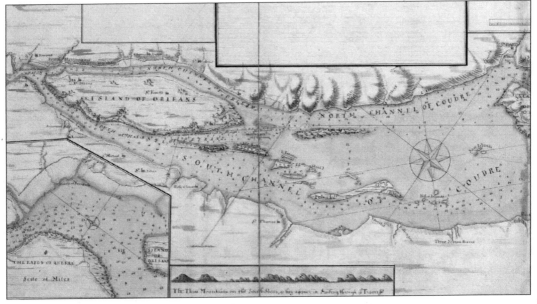

"A plan of the river St. Laurence from Green Island to Cape Carrouge," 1759, by James Cook

brought him to the attention of the Admiralty and the Royal Society. He was never to return to North America's Atlantic coastline. Instead, in 1768, he was given command of a former Whitby collier, renamed the *Endeavour*, and set off on his more famous adventure in the Southern Seas.

Canada's Arctic Dogsbody

Captain Bernier, 1853–1934

The voyages of Joseph-Elzéar Bernier to the Canadian Arctic were made with the typical mix of motives for such explorations: being "the first to get there," colonial expansion, trade, mapping, scientific discovery, and a dose of religion mixed in for good measure. But this time, the protagonist's efforts were continually thwarted by the fact that he was neither a British officer nor a member of the Anglo-Canadian ruling classes, but a French-Canadian.

Born to a sea-faring family in L'Islet, Quebec, Bernier was a cabin boy by the age of 14, and a captain of his own ship at age 17. All told he made 250 Atlantic crossings, 44 on wooden ships built at Quebec. His great dream, however, was to find the North Pole for Canada.

"I wanted to have the honour of reaching the pole for Canada," Bernier wrote, "and at the same time reaching the islands and giving them to Canada; and if any were discovered to annex them. They were ours; they were given to us by Britain on the 1st September, 1880." He presented his plans to the public and Canadian Prime Minister Wilfred Laurier repeatedly, but a North Pole expedition was never in the cards.

Laurier wanted to use Bernier to patrol the Arctic for Canada, annexing whatever he could. He ended up being the government's general dogsbody for everything that needed doing in the Arctic. From 1904 to 1925 Bernier made 12 trips to the Arctic and spent eight winters there.

Bernier threw himself into the job. He replenished provision caches and replaced messages left by the British before him. He set up cairns on the highest points, decorating them with caribou antlers, and he held patriotic religious ceremonies to stake Canada's claim, raising both flags and crosses. He set navigation markers in place, such as harbour beacons, and recorded topographical data to improve existing sea charts. He named newly annexed areas after his friends. He established new police stations and resupplied existing ones. He even transported prisoners. One Inuit prisoner, Nuqullaq, released from prison early because he had tuberculosis, was returned to his home

> One must observe nature, obey her as a master, and never try to conquer her.
>
> - Captain Joseph-Elzéar Bernier

community of Pond Inlet by Bernier, whereupon the disease spread throughout the northern coast of Baffin Island.

On his second expedition, 1908-1909, Bernier was accompanied by scientists, whom he supported to study meteorology, magnetic elements, minerals, fauna, flora, topography, and the thickness of moving ice. They had plenty of time for the latter, as the ship spent the winter trapped in the ice and they endured 90 days of darkness. Every day crew chopped ice out from around the hull to prevent the ship getting crushed or "nipped." They built a snow wall as shelter from the wind, and a windmill to supply them with electricity. For entertainment they held dances accompanied by fiddles and a pianola. They also had plenty of time for lucrative fur trading with the Inuit. Crew traded everything, including their uniforms. "The men are all naked, from the first mate right down the lowest of the waiters," Bernier grumbled. Some crew members also had sex with Inuit women while the Inuit men were away on hunting expeditions, with the result that many Nunavut Inuit children have French-Canadian ancestry.

Bernier celebrated the arrival of summer by proclaiming Canadian sovereignty over the entire Arctic archipelago as far as the North Pole.

He learned everything about surviving in the North from Inuit. Elders at Pond Inlet still speak of *Kapitaikallak* ("the short captain"), remembered as one of the rare visitors who treated their forefathers as equals and genuinely enjoyed their company. The crew donned traditional Inuit winter clothing—more effective against the cold—and depended on their expertise as guides. One of his crew members recalled asking an Inuk in each locality to sketch maps of the region's bays, rivers and sounds, which "they did with great readiness and ability... [with] full particulars of each place with the approximate distances and general directions and bearings." The crew held hunting competitions with Inuit, football matches and snowshoe races. They treated them to feasts so they understood that being Canadian was delicious: "They were now Canadian & therefore subject to our laws," wrote Bernier. "Well they could not see that, but I tell you they saw it when they came on board my vessel to a dinner to which I had invited them, and they had everything they wanted, and then they commenced to realize that it was a good thing to be Canadian."

This afternoon the captain played a little music for us, but I was unable to listen since I was making some prune jam.

- Crew member Ludger Lemieux

At his dinner parties Bernier also played records on his phonograph and sang favourite French-Canadian songs that Inuit on North Baffin Island still sing: "Alouette, gentille alouette," and "Il était un petit navire" ("there was a little boat"), a song that Inuit called *ilititaa*.

Pikey, Niviaqsarjuk, and Taptaqut, photographed by Bernier's expedition at Cape Fullerton (Qatiktalik), 1905.

On his third expedition in 1910, Bernier was told to patrol newly claimed Arctic islands, issue whaling licences to foreign whalers, and act as Justice of the Peace and protector of wildlife. This time he was also given *carte blanche* to attempt the Northwest Passage and head for the Pole—but ice had blocked the passage. Bernier made up for this disappointment by buying a whaling station and 960 acres of land near Pond Inlet from the government, becoming the only private land-owner in Baffin Island.

His last three trips to the Arctic were purely commercial, to trade with Inuit, hunt for gold and mine coal. To extract coal, he pulled up his ship beside a coal seam and the crew fired at it with guns, whereupon the coal rolled conveniently down the hillside to the ship.

Although Bernier had assured Canada's title to 740,000 square kilometres of the Arctic, and was a legendary hero in Quebec, it took Canada many years to recognize

Bernier's ship CGS *Arctic*, a wooden barquentine powered by steam and sail, ca. 1924.

his achievements. His once-celebrated ship *Arctic* rotted away. His tomb was vandalized and for years no-one knew where it was. Finally they named an icebreaker after him and put him on a stamp.

Nothing more human than a ship

The Canadian Arctic Expedition

The Canadian Arctic Expedition, 1913-1918, is more remembered for its disasters than its successes. In fact, it was not even very Canadian. Instigator Vilhjalmur Stefansson, an American anthropologist (Canadian-born), first obtained funding from the National Geographical Society (Washington) and the American Museum of Natural History and then asked Canadian Prime Minister Robert Borden for extra money to explore the "million or so square miles that is represented by white patches on our map, lying between Alaska and the North Pole." Worried that the American sponsors would give the US legal claim over any new land discovered, Borden told Stefansson he would fund the entire expedition, but only "on condition you become a British subject before leaving and the expedition [flies the] British flag." In addition to crew, the expedition employed 14 scientists from several countries and four Inuit guides and hunters.

The expedition's flagship, an ex-whaler called the *Karluk*, left Victoria, BC, in 1913. Its captain, Robert Bartlett, was from Newfoundland, and believed in placing science before exploration. He was an expert ice captain, and had sailed for the Arctic explorer Robert Peary.

Along the way the *Karluk* picked up four Iñupiat hunters. One hunter, Kurraluk, brought along his wife Qiruk, their two daughters and a cat. They also had a team of 16 dogs. Two other ships were to meet up with them in the Yukon.

The *Karluk* never made it. It got stuck in the ice. Instead of waiting around, Stefansson left to go caribou hunting and when he returned, the ship had drifted off. He gave up looking for it and departed to carry on with the expedition elsewhere, leaving

Kurraluk, Qiruk and children Qaualuk and Makpii, four of the survivors of the Canadian Arctic Expedition.

Bartlett and the rest of the crew to their fate. Kurraluk and another hunter, Kataktovik, kept the expedition supplied with fresh seal and walrus meat. But after drifting with the moving ice for four months, the ship was "nipped" in the ice and crushed to pieces.

Bartlett had already moved food, dog teams, and ammunition into a house built on the ice, and asked Qiruk to make them new boots and repair their clothes. Bartlett stayed on board until the last moment, playing Chopin's "Funeral March" loudly on the ship's phonograph. The *Karluk* sank within minutes of Bartlett stepping off, her yardarms snapping as she disappeared through a hole in the ice. "She creaked and groaned and, once or twice, actually sobbed as the water oozed through her seams," he recalled. "There is nothing more human than a ship in ice pressure."

Two parties of four departed—neither with Bartlett's full approval—and were never seen again. Bartlett and an Iñupiat hunter walked 700 miles to Siberia to get help. From there they got a lift to Alaska, where they went to find a rescue ship.

Those left behind fought bitterly as they waited. Two died and one was shot, possibly murdered. "The misery and desperation of our situation multiplied every weakness, every quirk of personality, every flaw in character, a thousandfold," reported an expedition member.

Billy was there ahead of us, they had no luck hunting. We started digging for some muskoxen carcus left here last fall, we dug all night and finally had the luck to strike one towards morning under an 8 foot snowbank, the wolfs had eaten one side, but there was still enough left for a feed for us and the dogs. It was very rotten but we did not wait, went down on hand and knees and dug in with our knives to fill our stomack.

- Karsten Andersen's diary, April 30, 1917, Melville Island.

Just before the next winter started, the remaining 14 survivors were rescued, as well as the cat. The cat had come on board at Esquimalt and had been adopted by fireman Fred Maurer. After the sinking of the *Karluk* she survived the ice marches and a period on Wrangel Island, and was rescued with the survivors in September 1914. The cat lived to a grand old age, producing numerous litters.

Only one survivor rejoined the expedition. A northern party, led by Stefansson, went on to explore the land, while a southern party carried out geographical, geological, biological and ethnological studies. The parties depended on hunting caribou, seal and muskox for food, dogs for travel, and word of mouth or letters passed from person to person for communicating.

In spite of the *Karluk* disaster, the expedition discovered four new islands (even Inuit weren't familiar with them), corrected maps of other areas, and collected thousands of specimens and artefacts. Anthropologist Diamond Jenness documented Inuit culture in great detail before other cultures influenced it. He recorded sounds and songs and collected samples of many aspects of Copper Inuit material culture. The expedition's biological, mineralogical and ethnological material was given to the National Museum of Canada.

The expedition was the Canadian government's first major exploration of the western Arctic, and the main means by which Canada came to know and control it. The expedition published its findings in 16 scientific volumes, and took 2700 metres of film footage and 4000 photographs. However, it remains a mystery as to why there was no official inquiry into Stefansson's abandonment of his crew, why no official film was made, and why much of the expedition's film footage was buried, lost and forgotten.

My thoughts went constantly,
To the great land my thoughts went constantly.
The game, bull caribou those,
Thinking of them I thought constantly.
My thoughts went constantly,
To the big ice my thoughts went constantly.
The game, bull caribou those,
Thinking of them my thoughts went constantly.
My thoughts went constantly,
To the dance-house my thoughts went constantly.
The dance-songs and the drum,
Thinking of them my thoughts went constantly.
I was beginning to waste away exceedingly from hunger.

- Song sung by Kaneucq, a Puivlik girl, at Prince Albert Sound, recorded by Roberts and Jenness.

MUKPIE · POINT BARROW ESKIMO GIRL ·
YOUNGEST PERSON ABOARD, ALSO ONE OF THE
SURVIVORS OF THE S.S.KARLUK ·
CANADIAN · ARCTIC (STEFANSSON) EXPEDITION ·

© LOMEN BROS.
NOME

Ruth Makpii Ipalook, one of the survivors of the expedition, lived until she was 97.

VI – EXHIBITING NATURE

Something had to be done with all the specimens brought home from these voyages.

Modern natural history grew out of collectors' attempts to name and order the contents of their specimen collections. At first, people took them home and showed them off to their friends. Eventually these personal collections migrated from people's homes into purpose-made buildings, and later still, they opened to the public.

The following chapters leap back in time to explore the birth and evolution of museums in Europe and the Canadas.

Officer's Room in Montreal, by Cornelius Kreighoff, 1846.

Les simples curieux

The birth of the great museums

The empty spaces of North America, Asia and Africa turned out to be full. Travellers brought back things no-one in Europe had ever seen before and gave them to their patrons, who kept them in *cabinets de curiosités* ("cabinets of curiosities") or *Wunderkammern* ("wonder rooms") in their private dwellings. Doctors, pharmacists, and even churches collected curiosities—several fifteenth-century churches had crocodiles hanging from their ceilings. Nature was seen as the busy and imaginative lieutenant under the command of Divine Providence.

Detail of Ole Worm's *cabinet of curiosities*, frontispiece of the catalogue for his museum, *Musei Wormiani Historia*, 1655. Ole Worm, a Danish physician, brought down objects and put them on the table in the middle for visitors "to touch with their own hands and to see with their own eyes, so that they may themselves judge how that which is said fits with the things."

Many "marvel mongers" sold their findings to the royalty and nobles of the sixteenth century, for whom their cabinet (the word meant room, not cupboard) was their status symbol: my collection is bigger than yours, my ships are faster, my explorers more adventurous, and my empire bigger. In addition to their scientific worth, cabinets were a promise of future riches, just as forests are seen today as potential for lumber or grazing land.

New collections of objects from afar were also a natural extension of the collecting of texts that explained the world. The great encyclopedic books of the Enlightenment, for example, especially the Comte de Buffon's *Histoire naturelle* (1749-88) and the 28-volume *Encyclopédie* by Diderot and d'Alembert (1751-72), had a wide readership. Diderot's encyclopedia gathered scattered information and systematized it in 74,000 articles. His ambition was Wikipedian—to bring together all that was known, about every field, to provide an understanding of the world based on reason and critical thinking.

How the contents of these nascent museums were organized was largely a matter of individual expression. Early cabinets were a glorious jumble of unrelated objects. Or objects might be divided into natural objects (*naturalia*) and those created or changed by humans (*artificalia*). For the *naturalia*, Linnaeus' divisions became the most common organizing principle: animal, vegetable, or mineral. Different cases held different kinds of artefact: rounded glass cases held stuffed birds, aquaria held river- and sea-creatures, jars held pickled specimens, and glass-fronted boxes held butterfly collections. Some shelves were left open so that people could pick things up and look at them.

> And on the pavement lay...
> Huge Ammonites, and the first bones of Time;
> And on the tables every clime and age
> Jumbled together; celts and calumets,
> Claymore and snowshoe, toys in lava, fans
> Of sandal, amber, ancient rosaries,
> Laborious orient ivory sphere in sphere.
>
> - Alfred, Lord Tennyson's description of an informal museum in *The Princess*.

But most prized of all were natural objects that seemed to straddle the categories—or monsters. Monsters were seen as omens: when Italian naturalist and collector Ulisse Aldrovandi acquired a "dragon" in 1572, it was seen as foretelling the outcome of the upcoming papal election.

Unlike Montaigne (1533–1592), who saw curiosity about the natural world as barbarous, Francis Bacon (1561–1626) believed that a constant back and forth between studying the particular and the universal was a necessary corrective to the abstract theories of natural philosophers. The study of nature's anomalies, he thought, might disprove their annoying axioms. But even the scientific academies couldn't resist curiosities for their own sake, and encouraged collectors to bring them back the weirdest things they could find.

Monster cockerel, in Ulisse Aldrovandi's *Monstrorum Historia* (History of Monsters), 1642.

Curiosity was a quality that crossed class boundaries—everyone has the capacity to marvel. The French distinguished between *les simples curieux* and the *vrais connoisseurs*: the former being the lower classes and private citizens who collected and displayed unusual objects for their exotic appeal, the latter being the naturalists who collected for scientific purposes and documented curiosities in academies. The London aristocracy sneered at the "knicknackatory" of the voracious collector Hans Sloane, whose collection formed the basis of the British Museum in 1753. His collection included a one-eyed pig, a silver penis protector, a vomit-inducing stick from India, a shoe made from human skin, and a manatee-hide whip for beating slaves. As natural and ethnographic oddities flowed in to Great Britain from the far reaches of the empire, the Royal Society of London urged that such curiosities could be used for both learning *and* education, or as "learned entertainment."

In Aldrovandi's collection in sixteenth-century Italy, public dissections were performed on pretty well every specimen that came through the door. Anatomical dissection as

La curiosité est vicieuse partout.

- Montaigne

experimental science was excellent theatre. As a museum tradition it was still alive at the Literary and Historical Society of Quebec as late as 1864. Dr. James Douglas, noted for his prowess in the operating theatre (he could amputate a leg in less than a minute), displayed a mummy he had looted from an Egyptian tomb and, in a spirit of participation, allowed visitors to snip off a lock of its hair and take it home.

Cabinets of curiosities became so big they needed their own buildings. In Italy, where the collecting of natural objects became a passion before anywhere else in Western

Europe, the Capitolines Museum opened in 1471 and the Vatican Museum in 1506.

Most early museums were private, but a London gardener and collector called John Tradescant (c.1570–1638) opened his home-grown collection to anyone with sixpence to spare. Some visitors were scandalized. "Even the women are allowed up here for six[pence]," wrote a German visitor, "[they] run here and there, grabbing at everything." This collection became Britain's first public museum, the Ashmolean, in 1683. One of the first public museums in France was the Louvre, which opened to the *hoi polloi* during the French Revolution in 1793.

The British Museum opened to the public 250 years ago, having paid £20,000 for Hans Sloane's collections. Other start-up costs were raised with a lottery. As knowledge became more compartmentalized, objects were organized not just into different cabinets but into different buildings. The British Museum's natural history and geography collections migrated into their own building, the Museum of Natural History, in the 1880s, and the Library moved away a century later.

Signor Cardinal Barberini did me the honour of requesting that I perform a small anatomical study of a monster that had been presented to him, that is, a calf with two heads, which I did in my house [in his private museum] in the space of four hours in the presence of my scholars two days ago, observing everything and making illustrations.

- Giovanni Faber, 1624

Critics complain that taxonomists are now in short supply in museums, and that advertisers and managers have replaced eccentric scientists devoted exclusively to a single sub-species. One palaeontologist at the Natural History Museum recalled that "the whale man" used to hide bottles of whisky in sheaths of blubber, and often fell into his pit of caustic chemicals and bones. Another researcher wove rugs on a loom she kept in her office. The museum keepers once lived on site, their children playing cricket on the lawns after closing time.

Historian Ken Arnold writes that museums tend to focus on one of three functions:

Monftrorum Hiftoria. 423

VII. Capitis vitulini geminati duplex effigies.

Calves with two heads. Engraving by Jean-Baptiste Coriolan. In Ulisse Aldrovandi's *Monstrorum Historia*, 1642.

narrative, i.e. telling a story; function, i.e. focusing on what objects are for; or taxonomy, i.e. organizing—the least entertaining of the three. He claims that taxonomy, plodding onwards in its exhausting inventory of life, took over until museums "classified themselves and their precious objects almost to extinction."

Otis and the safety elevator

World's Fairs

Museums were not the only showcases for the exotica of empire. The "Age of Exhibitions" started with Britain's Great Exhibition in 1851 and lasted for over a century. Referred to as universal exhibitions, *expositions universelles* or world's fairs, they took museums a step further in showing off the raw materials, skills and technologies of entire nations and empires—and provided entertainment at the same time.

More ambitiously still, they aimed to foster world peace, while competing in a friendly way. To be sure the Great Exhibition stayed nice and friendly, the Duke of Wellington ordered seven infantry battalions to stand by just in case.

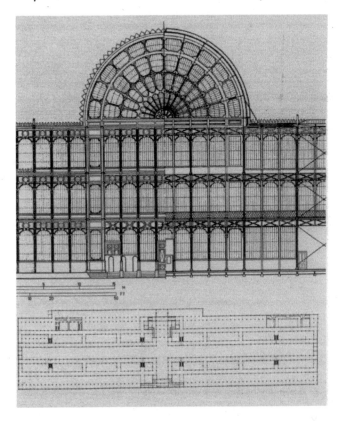

For Canada's contribution to the Great Exhibition of 1851, geologist William Logan used maps, narratives and carefully arranged specimens collected by the Geological Survey of Canada (see page 89) to create a powerful exhibit of scientific progress and the riches of a land open to immigrants. The exhibit ranked first among the colonies. Canadian and British papers praised its aesthetics and its message of a "modern" Canada open for business. When Logan took the exhibit to the Paris

Plan for London's Crystal Palace by gardener-architect Joseph Paxton, 1851

Exposition universelle in 1855, it was greeted with similar acclaim. By 1856, he had been knighted by Queen Victoria and made a Chevalier de la Légion d'Honneur by Napoleon III. He had set a pattern of display for museums, and created a new collective sense of self-confidence and self-respect among Canadians.

For the Great Exhibition of 1851, the Crystal Palace was constructed in London's Hyde Park over the top of ten elm trees. Built in four months and pulled down in three, it was the first ever pre-fab building, and the largest enclosed space on earth. Over 141 days, more than six million people (one-fifth of the British population) came to see more than 100,000 exhibits from Britain, the US and the British Empire. Exhibits ranged from a machine that could turn out between 80 and 100 cigarettes a minute, to a "silent alarm bedstead" which tipped the sleeper into a bath of cold water, and a large silver nose, its purpose unexplained.

Hundreds of tracts and books were published at the time to help people interpret the exhibition in light of their faith. Some feared the Exhibition would encourage materialism and pride; others saw it as heralding the New Jerusalem. Prince Albert, Queen Victoria's husband, made sure the first verse of Psalm 24 was on the cover of the program: "The earth is the Lord's and all that is therein; the compass of the world and they that dwell therein."

The Great Exhibition inspired other nations to hold their own universal exhibitions. They became the new mass medium. The Paris Exposition in 1900 attracted 50 million visitors, more than the entire French population—a record only broken at Expo '67 in Montreal. No other visual mass media reached so many before television. In total, 300 international large-scale exhibitions were held worldwide between 1851 and 2001, more than half in Europe, but also Sydney and Melbourne, Delhi, Calcutta, Hanoï, Montreal, New York and Chicago.

As to [the Vienna World Fair's] contribution to world peace, even the most simple-minded among us know what nonsensical chatter this is ... What we are doing here will be completely forgotten in five years.

- German engineer Max Eyth, 1873

At the New York World's Fair in 1853, Elisha Otis demonstrated a new safety elevator. In Paris at the Universal Exposition in 1855, the Tripp brothers of Canada exhibited a pile of gravel and bitumen (asphalt) which, dull as it looked, won them a contract to pave the streets of Paris (they went bankrupt before it could be delivered). At the Philadelphia Exposition of 1876, Alexander Graham Bell introduced the telephone, and in Paris in 1889, Thomas Edison gave his first public demonstration of the phonograph. The St. Louis World's Fair of 1904 introduced the safety razor, the ice cream cone, iced tea, and rayon.

The Prince of Wales as Chief Morning Star, carved in butter, at the Canadian Pavilion, British Empire Exhibition, Wembley 1925.

In addition to inventions, entire industrial systems were put on display. At the 1876 Philadelphia Exposition, a 2520-horsepower steam engine, 40 feet high, ran all the machinery in the hall. Visitors could also see, hear and smell steam fire engines, steam trains and steam pumps.

Universal exhibitions were "momentary centres of world civilization" which "assembled the products of the entire world in a confined space as if in a single picture," said German sociologist Georg Simmel in 1896. Canada created a Government Exhibition Commission in 1901, and participated eagerly in each. At the British Empire Exhibition in 1924-5, it showed off its railways and the Group of Seven artists.

Exhibitors at world's fairs increasingly emphasized their imperial power by including their colonized peoples as exhibits. The 1894 World's Fair in Antwerp, Belgium, contained a "human zoo" of 144 Congolese imported to Belgium by boat. Seven eventually died, and were buried in a mass grave. An exhibit of Koreans as cannibals at the Osaka Exhibition in 1903 helped Japan justify its subsequent colonization of Korea. The Turin exhibition of 1911 showcased the Italian Empire by recreating an Eritrean and Somali village with people dressed in traditional attire.

Canadian officials, however, worried that exhibiting Indigenous culture would contradict the image of Canada as a civilized nation. At the 1886 Colonial and Indian Exhibition in South Kensington, space devoted to Canada's Indigenous cultures was the size of "a moderately-sized dinner table," according to a disappointed anthropologist. In the 1925 British Empire Exhibition, the only reference to Indigenous peoples in Canada's exhibit was

Exhibition fatigue in the summer of 1900 - *Encyclopédie du siècle*, vol. 2.

a sculpture of Edward, Prince of Wales, dressed as an Assiniboine chief and made entirely of butter. The sculpture's dual purpose was presumably to emphasize Canada's dairy industry and its loyalty to empire.

As early as the 1880s, Germans were complaining of *Ausstellungsmüdigkeit*, or exhibition fatigue. Egbert Hoyer calculated that for the Paris Exposition of 1878, if someone planned to visit all 53,000 exhibitors for one minute each it would take them six months.

Still, universal exhibitions did not lose their appeal until the mid-twentieth century. Art biennials, such as that in Venice, are still going strong, but all that remains of the world's fairs now are a few architectural remains, including the Eiffel Tower and the Musée d'Orsay in Paris, and Montreal's Biosphere.

The calf with two heads

The first natural history museums in Quebec

In the early nineteenth century, natural history was the fashion for the professional bourgeoisie in Quebec. As in Europe, lawyers, doctors, engineers and merchants had private *cabinets de curiosité* and natural history museums in their homes. In Quebec, Judge Louis-David Roy had a collection of indigenous plants from the Chicoutimi and Malbaie areas, and lawyer James MacPherson Le Moine (see page 43) had a large bird collection, as did artist Cornelius Krieghoff.

Museums as public entertainment in Quebec began in 1824, when an innkeeper Tommaso Delvecchio, one of the first Italian immigrants in Quebec, opened his Museo Italiano in Place du Vieux-Marché in Montreal. As well as slaking a growing thirst for knowledge

> The study of nature, from a Christian point of view, is not only useful for things of this world, but raises us up towards God by speaking of his power with evidence and eloquence.
>
> - Louis-François Laflèche, Bishop of Trois-Rivières, 1885

about natural history, early museums in Quebec as elsewhere satisfied a popular taste for gawking at natural deformities and depictions of dramatically violent moments in history. Delvecchio exhibited a crocodile, a panther, a white bear with six feet, a lamb with eight legs, and a wax sculpture of assassin Charlotte Corday stabbing the French Revolutionary doctor Jean Paul Marat in 1793. In 1827 the Natural History Society of Montreal inaugurated its own museum.

In Quebec, Pierre Chasseur, a wood sculptor, gilder, and accomplished taxidermist, opened a private museum in 1826. It included 75 stuffed mammals, 40 reptiles and fish and 500 birds. It contained fossils, horseshoe crabs, Indigenous artefacts—and the axe used by a famous Montreal murderer to kill his pregnant wife. In spite of receiving government loans to keep his museum going, it was expropriated in 1836. The following

Ticket for the Musée Chasseur, 1831. Engraved by James Smillie, Quebec.

year Chasseur was arrested and imprisoned for his involvement with the Patriote Party. The Party held meetings in his house, by now emptied of its natural history specimens. In 1841 the government added Chasseur's collection to the natural history museum of the Literary and Historical Society of Quebec, located in Parliament House (see page 38)—but Parliament House burned down in 1854 and all was lost.

Public interest in museums was clearly not enough to sustain them financially, as is still the case today, but they had social as well as commercial purposes. Going to the museum was a sign of social distinction, showing you had good taste, education and the leisure to pursue scientific interests.

After Confederation in 1867, the museum movement spread across Canada, funded by a blend of private and government support. William Logan, geologist of the Province of Canada, had a large and carefully catalogued collection of specimens by 1845. After its success in London at the Great Exhibition of 1851, he was inspired to start a permanent geological museum in Canada. Part of the Geological Survey of Canada, it was the first truly public museum in Canada.

Museums also had an educational vocation. Abbé Jérôme Demers had opened a museum in the Séminaire de Québec in 1806, and in mid-century, the new universities of McGill and Université Laval also opened natural history museums for teaching purposes. Private classical colleges did the same. Religious congregations, including the Soeurs de la Charité and the Soeurs du Bon Pasteur de Québec as well as Mechanics Institutes all had their own museums. By the late nineteenth century no self-respecting educational institution was without its museum.

The calf with two heads. Collections Université Laval.

The hunger for specimens was voracious. Collectors in the field could be upper-class amateur botanists such as Lady Dalhousie, or parish priests such as Léon Provancher, but they could also be fishermen, hunters or trappers. From 1859-1866, over 11,000 zoological specimens were sent to the Smithsonian Museum in Washington by fur traders and trappers working for the Hudson's Bay Company who had been recruited as collectors by the museum scientists.

The character of the collections often depended on their curators. At Université Laval, priest and geologist Joseph-Clovis-Kemner Laflamme continually rearranged the fossil collections as scientific theories evolved, and organized exchanges with the Smithsonian in Washington and the Muséum d'histoire naturelle de Paris. By 1875, Université Laval's museums had amassed 1300 birds, 100 mammals and more than 12,000 insects. At the Collège Saint-Laurent, idiosyncratic curator R. P. Joseph-C. Carrier's collection included beetles and shells but also 1500 buttons, as well as a "département d'indianeries."

The sense of wonder at curiosities gradually fell into disrepute. All nature's marvels were in principle explicable, and scientific knowledge was associated more with diligence and discipline than with shock and delight. Natural history split into numerous subspecialties and field studies, and studying live creatures became more interesting than looking at stuffed dead ones. By the 1890s, natural history museums were no longer a *sine qua non* of educational institutions, and natural history was no longer in the baccalaureat exams.

Brother Florian Crête, who was deaf, was determined to restore this sense of wonder. In 1915 he became director of Montreal's Musée d'histoire naturelle de l'Institution des Sourds-Muets, founded in 1882. Like the keepers of the Renaissance *cabinets de curiosités* with their open shelves and handling tables, he believed the sensory experience of objects was as important to education as the intellect. He let visitors touch and manipulate some of the museum's 13,000 specimens, and created educational games. "Whatever kind of museum we work in," he wrote, "if we do not have or cannot communicate to our visitors warmth, affection, comprehension, or the sense of wonder … our museum is only half alive, distant, draped in erudition as if in a cloak of indifference."

Sinking into the mud

The Geological Survey and the Canadian Museum of Nature

The Union of Upper and Lower Canada into the Province of Canada (today's Ontario and Quebec) in 1841 was deeply contentious. Lord Durham hoped that economic development would "do more for the present pacification of the Canadas than anything else." The government needed to know what natural resources were available, but also feared it might get a raw deal in border negotiations with the US if it didn't know what was at stake. All it had was a patchwork of surveys, mostly by British officers. Proper geologists were the people for the job.

The French-speaking majority in Lower Canada thought that public surveys under a British regime smacked of foreign capitalist interference. However, insistent petitions

Chic-Choc Mountains, by Sir William Logan, Gaspé journal, 1843-44.

to the government from the Natural History Society of Montreal and the Literary and Historical Society of Quebec led to the launch of the Geological Survey of Canada in 1842. The Survey was to make "a full and scientific description of the country's rocks, soils, and minerals, to prepare maps, diagrams, and drawings, and to collect specimens to illustrate the occurrences." This is more or less what it does to this day.

Its main mission, at the beginning, was to find coal. Its first director, Montreal-born William E. Logan (1798 –1875), became interested in geology while managing his family's coal operations in Wales, and saw the need for accurate maps to locate the seams. Logan set off to survey the country.

Logan walked for hundreds of kilometres, counting his steps to measure distance. He worked from dawn to dusk, then wrote notes, sketched and drew maps by the light of the campfire at night. Logan describes a summer travelling along the St. Lawrence River, "sleeping on the beach in a blanket sack with my feet to the fire, seldom taking my clothes off, eating salt pork and ship's biscuit, occasionally tormented by mosquitoes." When he was accompanied by his multilingual Mi'kmaq companion, John Basque, the food was much better. John Basque was not only an expert canoeist, but an excellent hunter and cook. He cooked birds, fish and mammals, including porcupine, flavouring his dishes with local plants such as wild chives.

Logan quickly gave up on finding accessible coal in the newly named Province of Canada—the rocks were too old. He began to look for other useful rocks and minerals instead, and found rich beds of copper, nickel, uranium, iron, platinum, and other metals in the ancient rocks of the Precambria.

When locals saw Logan at work in the field, with his wild red beard, dressed in patched clothes and torn boots, his hair matted with spruce gum and his glasses cracked, muttering continuously,

Excerpt of Sir William Logan's diary for 1845, listing his Indigenous guides, Mathias Sanōracé, Michael Teauāhqua, Enias Atawāk-hun and Thomas Carahō-to, whom he paid four pounds per month plus a pair of mocassins.

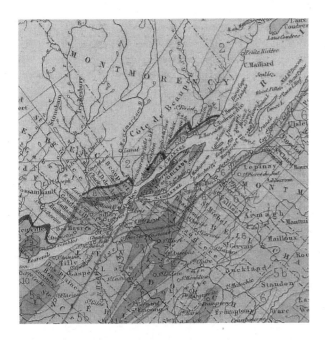

Detail of William Logan's 1865 "Great Map" showing Logan's Line, the fault he located between the St. Lawrence Lowlands to the north and the folded and overturned rocks of the Appalachians to the south.

hammering at rocks and carefully saving the chips, they thought him completely mad.

In the early years of the Geological Survey of Canada, crews followed Logan's example, doing most of their travelling on foot, in dogsleds or boats. Indigenous guides showed them the trails and portages as well as canoeing, hunting, cooking, carrying the supplies, and building new canoes when needed. Facing threats of forest fires, drowning, and bears, they climbed mountains, crossed rivers, struggled through dense bush, sheltered in the open and ate whatever was available.

This is how the main geological formations of the country were mapped. In the city of Quebec, Logan used fossils to locate the fault dividing the St. Lawrence lowlands (younger) from the Appalachians (older), known today as "Logan's Line." It was Logan who identified the ancient Precambrian Laurentian strata in the Canadian Shield, dating from, in his own words, "God knows when." In spite of British objections, Logan insisted that his discovery be named after the Saint Lawrence River.

Canada was cutting the colonial apron strings. In 1841, Logan had sent two crates of fossils to London for identification, and heard nothing. Ten years later, while at the Great Exhibition, he investigated further and was "mortified" to find they were still unopened in the basement of the British Geological Survey. He brought them home and appointed a palaeontologist, Elkanah Billings, to his own staff.

Logan's 1865 "great map" included the Maritimes as well as the Province of Canada. By Confederation in 1867, the map had normalized the idea of the new country. His friend and fellow geologist William Dawson (see page 92) saw Confederation as an accurate political reflection of the country's geology: the Laurentian strata stretching across the continent showed that "local subdivisions [are] arbitrary and artificial."

The Survey was now permanent, and Confederation massively increased the area it had to cover. Alfred Selwyn, director from 1869 to 1895, surveyed along the course of the new Canadian Pacific Railway, British Columbia's reward for joining Confederation. He lost his precious field notes on the geology of the Albreda Pass when his horse ate them.

George M. Dawson (William's son), Survey director from 1895 to 1901, explored the Yukon, preparing the way for the Klondike gold rush. Dawson City, centre of the goldrush, was named after him. Meanwhile he sent 25-year-old Joseph Tyrrell, a junior employee, to take a look at Alberta. Instead of finding coal, he discovered dinosaur bones. The world-famous dinosaur museum, the Royal Tyrrell Museum in Alberta, was named after him.

George Dawson wrote extensively about the Haida, Kwakiutl and Secwépemc (Shuswap) nations and published a linguistic map of British Columbia. He persuaded Canada to keep Indigenous artefacts instead of sending them to European museums, which were competing for them. Indigenous artefacts were some of the first in the collections of the Canadian Museum of History—which is now in the process of returning these cultural artefacts to their Indigenous owners.

Technologies changed with time. Photos replaced sketches. Airplanes and helicopters replaced walking, canoes, horses and dogsleds. From 1952 to 1958, the Survey mapped about half as much of Canada as had been mapped in the previous 110 years, mostly because of helicopters.

When Logan was first appointed to the Survey in 1842, he bought a few display cases with his own money to exhibit his rock collections in the Survey's headquarters, a "small and dark room" belonging to his brother in Montreal. The "Survey Museum," as people called it, moved with him into the Montreal house he rented for it in 1856, and expanded its collections to botany, mineralogy, palaeontology and zoology. In 1881, the Survey moved to Ottawa, the new capital. There the museum gained a grand Tudor Gothic style building named the Victoria Memorial Museum.

There has been voted in Canada a sum of £1500 to commence a geological survey of the province. It will be an arduous undertaking. In the Spring and Summer mosquitoes and black flies are a perfect torment in the woods, and in the woods the provincial geologist will have to spend the chief part of his time, as but a small part of the country is yet cleared. In addition to th[e] geological features of the Country he will have to ... make a map of the river and mountains. No correct one exists.

- William Logan, October 19, 1841

Memorandum from Great Whale River: Three Boxes specimens etc. addressed 'R. Bell. Montreal' To go to Moose Factory. The following – to remain at Great Whale River for our return: 1 keg with pork – 2 Boxes with sundries – 1 Bag (full) flour – 2 Hams – 1 Water barrel – 1 shot gun. Will Mr Spencer kindly get 'Crow' and his friends to draw carefully a sketch map of coast of H[udson] Bay & canoe route all way to Fort Chimo. Also a sketch map of Great Whale River and if possible Little Whale River.

- Geologist Robert Bell, The Geological Survey of Canada, 1877

Victoria Memorial Museum, Ottawa, 1911.

Three hundred Scottish stonemasons were imported to help build this museum, which also housed the National Gallery. But the building's most important vocation was illustrated by the carved stone panels above the doors and windows depicting Canadian flora and fauna.

However, the geological surveying for this ambitious new building was clearly not up to par: the foundations soon began to shift. One day bricks and stones began shooting out from the walls, hitting and injuring some construction workers. It was sinking into the mud, and the tower had to be removed in 1915. When Parliament was damaged by fire in 1916, the museum had to vacate the premises for the evicted MPs to move in. The Senate occupied the Fossil Gallery for four years.

The history of the Ottawa museums reads like moves on a gameboard. In 1956, the Victoria Memorial Museum divided up its natural history and ethnographic collections into the National Museum of Natural Sciences and the National Museum of Man (later the Museum of Civilization). The Canadian Geological Survey moved to Booth Street, and the National Gallery of Canada to its new quarters downtown. The Museum of Civilization (now the Canadian Museum of History) moved to Gatineau. By 1990, the newly named Canadian Museum of Nature had the old towerless building all to itself. But there was still not enough room for its over ten million specimens. In 1997 most of its collections moved to the museum's new research and collection facility in Gatineau.

Slate pencils

William Dawson and the Redpath Museum in Montreal

Sir William Dawson (1820–1899) was principal of McGill University from 1855–1893 and one of Canada's most influential nineteenth-century geologists. His monumental body of work on geology and palaeobotany was based on 50 years of painstaking field research.

He first discovered fossil plants as a schoolboy in Pictou, Nova Scotia, while he was digging out rock behind the schoolhouse to make slate pencils. He was fascinated by the "college cabinet" at the Pictou Academy, and learned to prepare birds and moths for his own mini-museum in a cupboard at home. These childhood specimens later formed the basis of his teaching collections at McGill University, and some specimens from the

Diagram of the earth's history, in William Dawson's *The Story of the Earth and Man*, 1873

Pictou cabinet were later donated to the Redpath Museum. When Charles Lyell, the father of modern geology, visited Pictou in 1842 to examine its coal deposits, his guide was Dawson, who became his lifelong protégé, confidant, and disciple. Together they discovered the earliest reptile remains then known in North America, which were of importance to palaeontologists and later to evolutionists.

Dawson was educated at Edinburgh University, but after being refused the chair in natural history there because he was "a mere colonist," he accepted a job as principal of McGill. On arrival, he found the university consisted of two "unfinished and partly ruinous" buildings standing in a cow pasture. A winding cart-track, almost impassable at night, was the only access to town.

Dawson's creative research in geology, his creation of programs in biological and physical sciences as well as engineering and commerce, and his belief in higher education for women, changed McGill's reputation for all time.

When he arrived, the university's natural history collection consisted of a single fossil. A new wing of the Arts building included a room for a small natural history museum—the third in Montreal after the Natural History Society Museum and Logan's Geological Survey of Canada. Dawson collected specimens on his summer holidays, and cultivated wealthy benefactors such as William Molson (they went bison hunting together) to help buy more. He gave his treasured *Archeopteris gaspiensis*, a 280 million-year-old fossilized frond that he had found in the Gaspé and had discussed with Darwin, to the new museum. By 1862, the tiny museum had 10,000 specimens and was so overcrowded he had to take some home.

While at Edinburgh University, Dawson had spent many hours in its museum, a series of top-lit halls in a Greek-inspired quadrangle designed by Scottish architect Robert

Adam. Dawson dreamed of having a similar self-contained museum at McGill, with a scientifically arranged collection to represent the extent and diversity of God's creation, and a lecture hall. He felt that a proper museum was not only necessary for illustrating lecture topics, but was essential to the growth and scientific importance of McGill. When he was offered a job at Princeton University, sugar baron Peter Redpath said he would build Dawson a museum if he would stay. Completed in 1883, the Redpath is the oldest purpose-built museum in Canada. Its architect was the self-taught Alexander Cowper Hutchinson, son of Scottish immigrants like Dawson. Hutchinson had learned stone-cutting when he was 12, and was comfortable combining architectural styles from any period. Dawson insisted on a neoclassical "Grecian" look like Edinburgh's museum, with a great central hall flooded with light in which would hover a cast of the British Museum's *Megatherium* skeleton.

We have the freedom and freshness of a youthful nationality. We can trace out new paths which must be followed by our successors; we have the right to plant wherever we please the trees under whose shade they will sit. The independence which we thus enjoy, and the originality which we can claim, are in themselves privileges, but privileges carry great responsibilities.

- William Dawson, Presidential address to the Royal Society of Canada, 1882

Fossil specimens were arranged in order of geological time. Dawson's son George Mercer Dawson (see page 92), inspired by provincial museums in France, encouraged his father to include "a small typical local collection, with a map" to fuel local interest in natural history, in addition to collecting prestigious international specimens. Learned associations from America and Britain met in the new lecture theatre, and the Ladies Educational Association heard lectures on botany, zoology and "the geology of the Bible Lands."

O Lord, how Manifold are Thy works! All of them in wisdom Thou hast made.

- Plaque in Redpath Museum entrance hall

But the museum constantly had money problems, as do most museums, and the board often had to dig into their pockets. Dawson persuaded students to volunteer in the various departments. Curator Thomas Curry was assisted by Paul Kuetzing, a piano factory employee, who mounted and labelled skeletons in his spare time. Edward Ardley, the janitor, rose in the ranks to become assistant curator, while his son continued to shovel the snow and stoke the furnace.

By 1897 however, the museum couldn't afford to buy jars to hold the specimens. Once one of the top six museums in North America, by the early twentieth century the museum was in a state of neglect. The strong personalities that built and sustained it had died, and the new guard were less personally and financially invested.

Redpath Museum, ca. 1910

Renovations in the 1950s included new lighting on the displays. "For the first time since the gas jets had been turned off in the 1920s," said one regular visitor, "it was possible to enjoy the museum without a flashlight on dark winter afternoons."

Now the museum's collections have a new poignancy: the lion, gorilla and wolf, juxtaposed with the specimens of extinct species such as the passenger pigeon and the dodo, remind visitors of endangered species and the mass extinction that is probably underway, the sixth in the past 300 million years.

Sentinel species

Natural history and the future of nature

The mapping, exploring, discovering, collecting and classifying of our natural history continues to this day. To identify animals, the twenty-first-century naturalist is as likely to work with a smartphone, a scanning electron microscope, or a bioacoustic echosounder as with a pair of binoculars. In addition to referring to their guidebooks to identify animals by their form or colour, they might also carry a pocket genome-sequencing kit to properly identify its "clade"—the group of organisms that includes its ancestor and its descendants (or its branch on the tree of life). Birds, it turns out, are in the same clade as dinosaurs.

Studies are carried out by researchers in government, university or privately funded research institutes and museums, but also by amateurs belonging to dozens of voluntary associations or working as individuals. Volunteers tag butterflies, monitor birds using their smartphones, or count spider crabs in underwater timelapse photos. Some citizen-science projects, such as eBird, combine a social-media interface with crowd-sourced identification. eBird is one of the world's largest biodiversity-related citizen-science projects, available in every country in more than 40 languages. eBirders around the world use their smartphones to contribute more than 100 million bird sightings each year, and the accumulated data are changing our understanding of

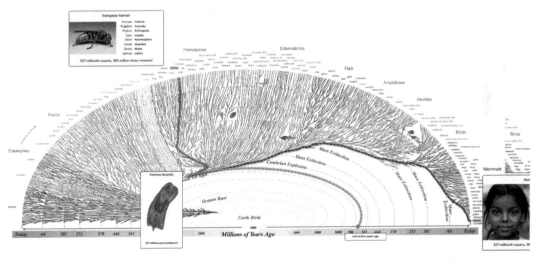

A hornet is our 327 millionth cousin, 303 million times removed. The online interactive Tree of Life Explorer, by Evogeneao, illustrates in simple form the core evolution princple that "we are related not only to every living thing, also to everything that has ever lived on earth." https://www.evogeneao.com/en/explore/tree-of-life-explorer

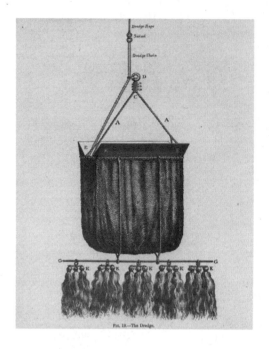

Fig. 18.—The Dredge.

Sampling from the seafloor: the HMS *Challenger* dredge (1872) used to collect shells of planktonic *Foraminifera* from the ocean floor. Modern ships are now comparing their own samples with those found by the *Challenger*.

bird migration, habitat use, distribution and abundance. Through another citizen-science research project called Zooniverse, people with mobility restrictions or living in city basements can identify distant galaxies in deep space, analyze cricket songs in Hawaii, or spot marine iguanas in the Galápagos Islands using remote camera data.

New technology is also being used to date the world all over again. Michel Lamothe studies the age of rocks using light—even better than carbon dating. Bioacoustic devices can record sound far beyond the range of human hearing, such as the songs of hatchling turtles calling out to each other through their shells to coordinate their birth times and thwart predators.

Bioacoustic devices are being used to help save species threatened by climate change or habitat loss. Right whales, for example, on the brink of extinction, began to move north around 2010, chased into busy northern shipping channels by climate change. By 2019, 50 of 400 remaining individuals had been killed, mostly by ship traffic and abandoned fishing gear. Bioacoustics researcher Kimberley Davies from New Brunswick and her team used a kind of underwater acoustic drone called a "glider" to detect whale calls and give their precise location to ships so they could avoid them—creating a kind of mobile marine sanctuary. The Canadian government fines ships that take no evasive action $250,000. In 2020 and 2021, there were no recorded right whale deaths in the Gulf of St. Lawrence due to ship strikes. Surveillance has its good sides.

Indigenous people are using modern technology to contribute data on wildlife in their traditional territories. With the Pineshish-Piyaasiis Boreal Bird Project, for example, Conrad André Kapesh from Mtimekush-La John, Eric Kapaé from Pessamit and Gloria Vollant from Uashat mak Mani-utenam are collaborating with the First Nations of Quebec and Labrador Sustainable Development Institute and Bruno Drolet from the Canadian Wildlife Service to monitor breeding birds in the forests of Northern Quebec using recording devices strapped to trees.

CCGS *Amundsen* collecting samples in Gibbs Fyord, Baffin Island, 2008. Photo Martin Fortier/ArcticNet

Some scientists are using collections made by early naturalists to track environmental change. They are comparing samples of "sentinel species," or organisms highly sensitive to environmental disturbances, that were collected long ago with samples taken today. Frédéric Raymond and Warwick Vincent (Université Laval) are working with scientist Anne Jungblut at the Natural History Museum in London on specimens collected by Captain Henry Feilden in 1876 during the British Arctic Expedition to the North Pole. Although his ships never made it to the Pole, they returned with many specimens, including a slimy blue-green algae, or Nostoc. This algae withstands very low temperatures and also fixes nitrogen, so it sustains food webs in Arctic freshwater and soil. By comparing this historical specimen with current and future specimens using genome sequencing, scientists can observe climate change and its effects. In another study, polar biologists

The swabs attached to the dredge were bristling with sea-urchins and star-fish, and its living contents showed that there is a great abundance of animal life in this part of the ocean.

- George S. Nares, *Narrative of a voyage to the Polar Sea during 1875-76 in H.M. ships 'Alert' and 'Discovery'*

Gabriele Treu and Mikkel-Holger Sinding and their colleagues are comparing Arctic wolf furs stored in museums from 1869 with present samples to chart mercury pollution.

New technologies are revealing millions more species than we knew existed. To identify species in the traditional way, intact specimens are needed to compare physical similarities (morphology). But taxonomy using genome sequencing requires only a bit of claw or a hair. Familiar names of frogs or butterflies turn out to be disguising more species (or "cryptic species"), each with its separate evolutionary heritage. DNA barcodes, or the instruction manual for making each species, can also show the history of their geographical distribution, and how a species becomes invasive.

This new way of classifying organisms has led to a "species problem," or the lack of an agreed definition of a species. Some propose abandoning the concept altogether. But for the time being, the Linnaean names and methods are still necessary as a common language linking the diverse disciplines: a "certain degree of fuzziness" is recommended. Also, genetic information alone can sometimes make distinct species seem deceptively similar.

Massive databanks of genetic information are the digital museums of the twenty-first century. BIOSCAN, an international database, has the ambition to collect barcodes of genetic information for 2.5 million species by 2026, and to achieve a "comprehensive understanding of multicellular life by 2045."

Many old museums are combining the old and new approaches to collecting, using their old specimens to create genetic databases. Kamal Khidas, curator at the Canadian Museum of Nature, is using specimens collected by naturalists over the years to create the National Biodiversity Cryobank of Canada, the first of its kind in the world. Its goal is to collect and preserve DNA samples from every species found in Canada "and preserve Canada's biodiversity for ever." DNA is being taken from plants, skins or bones collected for museums long ago or from fresh specimens to form a new kind of collection. Instead of in a drawer or a cupboard, samples are stored in tiny "cryovials" in cylindrical freezers. By 2018 the bank had collected 30,000 specimens. Inside one freezer were 7000 tissue samples of vertebrates submitted by Parks Canada, including grizzly bears and Blanding's turtles.

Old-fashioned species hunters without portable genome sequencing kits are not yet extinct, however. Stéphane Le Tirant is a founding member of the Montreal Insectarium, which studies and publishes on the insects in its collection. Le Tirant has been studying insects since he was five, and has created insectaria in Hong Kong, China and Newfoundland. Like the naturalists of old, he travels around the world discovering and naming new species. One was *Golofa limogesi*, a rhinoceros beetle which he named after René Limoges, a technician at the Insectarium. Rhinoceros beetles can be up to

17 centimetres long and can lift weights of two kilograms. He also studies *Phylliidae,* insects that look just like leaves, and imitate foliage in the wind by rocking as they walk. Le Tirant started up the Quebec branch of Monarch Watch, inviting amateur scientists to tag the declining population of monarch butterflies and plant or protect milkweed, which is where the females can stop to lay eggs on their multi-generation 5000-kilometre migration to Mexico.

* * *

While the vocabularies of biodiversity and environmental sciences have largely replaced the vocabulary of "natural history," the term still reminds us that our relationship with nature has both a history and a future. Like its history, the future of our relationship with the planet will combine the work and ideas of people from many nations and disciplines. It will involve wonder, surprise, and horrendous mistakes. It will also include art by a new generation of visual artists such as Mitsuko Kurashina, Yana Movchan and Bonnie Devine, and poetry, such as Ken Howe's collection *The Civic-Mindedness of Trees* and Bruce Taylor's *No End in Strangeness.*

If our planet is to survive, we'll need both the light touch of the monarch butterfly as it lands on a milkweed flower, and its submission to the demands of a journey whose purpose is not just individual, but multi-generational.

The sentinel raccoon. Moses Griffith, 1806.

FURTHER READING

I - NOVELTIES OF NATURE

Blühm, Andreas, and Louise Lippincott. (2005). *Fierce Friends: Artists and Animals, 1750-1900.* London: Merrell.

Chartrand, L., Duchesne, R., & Gingras, Y. (2008). *Histoire des sciences au Québec: De la Nouvelle France à nos jours.* Montréal: Boréal.

Desmond, R. (1995). *Kew: The History of the Royal Botanic Gardens.* London: Harvill Press & Royal Botanic Gardens.

Endersby, Jim. (2020). *Imperial Nature: Joseph Hooker and the Practices of Victorian Science.* University of Chicago Press.

Findlen, Paula. (1994). *Possessing Nature: Museums, Collecting, and Scientific Culture in Early Modern Italy.* University of California Press.

Foster, Janet. (1998). *Working for Wildlife: The Beginning of Preservation in Canada.* University of Toronto Press.

Gates, Barbara T. (2007). "Why Victorian Natural History?" *Victorian Literature and Culture,* Vol. 35, No. 2, pp. 539-549.

Jacques Rousseau. (2003). "Sarrazin, Michel," in *Dictionary of Canadian Biography,* vol. 2, University of Toronto/Université Laval. Retrieved from http://www.biographi.ca/en/bio/sarrazin_michel_2E.html.

Jarrell, Richard A. (2003), "Kalm, Pehr," in *Dictionary of Canadian Biography,* vol. 4, University of Toronto/Université Laval. Retrieved from http://www.biographi.ca/en/bio/kalm_pehr_4E.html

Lahaise, R. (1978). Voyage de Pehr Kalm au Canada en 1749. *Voix et Images,* 3(3), 487–490. Retrieved from https://doi.org/10.7202/200128ar.

Natural History Museum. "Kalm, Pehr (1716-1779)" at Plant Collectors, JSTOR Global Plants database. Retrieved from https://plants.jstor.org/stable/10.5555/al.ap.person.bm000152059.

Rieppel, O. (2011). "Species are individuals—the German tradition." *Cladistics,* 27: 629-645. Retrieved from https://doi.org/10.1111/j.1096-0031.2011.00356.x

Royal Museums Greenwich. (n.d.). "Longitude found - the story of Harrison's Clocks." Retrieved from https://www.rmg.co.uk/discover/explore/longitude-found-john-harrison.

Rudwick, Martin J.S. (2014). *Earth's Deep History: How It Was Discovered and Why It Matters.* University of Chicago Press.

Rundell, Katherine. (2022). "Consider the Hummingbird." *London Review of Books*, 3 November.

Secord, J.A. (2018). "Global geology and the tectonics of empire," in eds. Curry, H.A, Jardine, N., Secord, J.A., Spary, E.C. *Worlds of Natural History.* Cambridge University Press.

Wilcove D. S. and Eisner T. (2000). "The impending extinction of natural history." *Chronicle of Higher Education* 15: B24.

Wilkins, John S., Frank E. Zachos, and Pavlinov I. IA, eds. (2022). *Species Problems and Beyond: Contemporary Issues in Philosophy and Practice.* Boca Raton: CRC Press, Taylor & Francis Group.

II - INTERROGATING NATURE

Memory maps: Indigenous natural history

Aporta, C. (2004). "Routes, trails and tracks: Trail breaking among the Inuit of Igloolik." Études/Inuit/Studies, 28(2), 9–38. Retrieved from https:// doi.org/10.7202/013194ar.

Biggar, HP & Cook, Ramsay (eds.). (1993). *The Voyages of Jacques Cartier,* trans. HP Biggar. University of Toronto Press.

Champlain S. de & Thierry É. (2019). *Les œuvres complètes de Champlain.* Quebec: Septentrion. Œuvres.

Chartrand, L., Duchesne, R., & Gingras, Y. (2008). *Histoire des sciences au Québec: De la Nouvelle-France à nos jours.* Montreal: Boréal.

De Vorsey, Louis. (1992). "Silent Witnesses: Native American Maps." *The Georgia Review*, vol. 46, no. 4, pp. 709–726.

Delâge, Denys. (1995). *Bitter Feast: Amerindians and Europeans in Northeastern North America, 1600-64.* Vancouver: UBC Press.

Driver, Harold E. (1969). *Indians of North America.* University of Chicago Press.

Gagnon, F. M., Ouellet, R., Senior, N. (2011). *The* Codex canadensis *and the writings of Louis Nicolas.* McGill Queens University Press.

Knight, V. J. (2009). *The search for Mabila: The decisive battle between Hernando de Soto and Chief Tascalusa.* Tuscaloosa: University of Alabama Press.

Kuhnlein H & Turner N. (1991) "Traditional Plant Foods of Canadian Indigenous Peoples." FAO. Retrieved from http://www.fao.org/wairdocs/other/ai215e/AI215E05.htm.

Lahontan, Baron. (1709). *Nouveaux Voyages de Baron Lahontan.* La Haye: Chez les Freres L'Honoré. p 105. Retrieved from https://archive.org/details/McGillLibrary-rbsc_lc_nouveaux-voyages-de-baron-lahontan_lande00500-17820.

Le Clercq, Chrestien. (1691). *Nouvelle relation de la Gaspésie: Qui contient les moeurs & la religion des sauvages gaspésiens Porte-Croix, adorateurs du soleil, & d'autres peuples de l'Amérique septentrionale, dite le Canada: dédiée a Madame la princesse d'Épinoy.* Paris: Chez Amable Auroy. Retrieved from https://archive.org/details/nouvellerelationoolecl.

Lopez, Barry. (1986). *Arctic. Imagination and Desire in a Northern Landscape.* New York: Scribner's.

MacDonald, J. (2018). "Planets in Inuit Astronomy." *Oxford Research Encyclopedia of Planetary Science.* Retrieved from http://oxfordre.com/planetaryscience/view/10.1093/acrefore/9780190647926.001.0001/acrefore-9780190647926-e-59.

Miller, J. R. (2018). *Skyscrapers Hide the Heavens: A History of Native-Newcomer Relations in Canada.* Toronto: University of Toronto Press.

Moodie, D. (2021). "Cartography in Canada: Indigenous Mapmaking." In *The Canadian Encyclopedia.* Retrieved from https://www.thecanadianencyclopedia.ca/en/article/cartography-in-canada-indigenous-mapmaking.

Northcott, H. C., & Wilson, D. M. (2008). *Dying and Death in Canada.* Peterborough, Ont: Broadview Press.

Palomino, J.-F. (2012). "Cartographier la terre des païens : la géographie des missionnaires jésuites en Nouvelle-France au XVIIe siècle." *Revue de Bibliothèque et Archives nationales du Québec*(4), 6-19. doi:https://doi.org/10.7202/1012093ar

Simon, Lorène. (2015). "Échanges entre des religieuses de l'Hôtel-Dieu de Québec et un apothicaire de Dieppe au XVIIIe siècle." *Pharmacopolis: Revue québécoise d'histoire de la pharmacie*: 2.

Vogel, Virgil J. (1987). "The Blackout of Native American Cultural Achievements." *American Indian Quarterly*, vol. 11, no. 1, pp. 11–35.

More fun than embroidery: Lady botanists of Quebec

Blair, L. (2016). "Moose in Flames: The Story of the Literary and Historical Society of Quebec." In Blair, L.; Donovan P., & Fyson, D. *Iron Bars and Bookshelves: A History of the Morrin Centre.* Montreal: Baraka Books.

Reid, Deborah. (2015). "Unsung Heroines of Horticulture: Scottish Gardening Women, 1800 to 1930." PhD thesis, University of Edinburgh.

Gallichan, G. (2011). "La bibliothèque personnelle du gouverneur Dalhousie." *Les Cahiers des dix*, (65), 75–116. https://doi.org/10.7202/1007772.

Graham, Robert. (1839). "Extracts from a Report on the Progress and State of Botany in Britain, from March 1838 to February 1839 [...]." Third Annual Report and Proceedings of the Botanical Society (1838-39).

Hardy, Suzanne. (1999). "Trois grandes dames botanistes à Sillery au XIXe siècle." *La Charcotte*, vol 13, n° 1.

Le Moine, J. M. (1882). *Picturesque Quebec: a Sequel to Quebec Past and Present*. Montreal: Dawson.

Matthews, C.C. (2013). "Women writers and the study of natural history in nineteenth-century Canada. " PhD thesis, University of British Columbia.

Savard, Pierre. "Sheppard, William." In *Dictionary of Canadian Biography*, vol. 9, University of Toronto/Université Laval, 2003–, Retrieved from http://www.biographi.ca/en/bio/sheppard_william_9E.html.

Pringle, James S. (1985). "Anne Mary Perceval (1790-1876): An Early Botanical Collector in Lower Canada." *Canadian Horticultural History*, vol 1, n° 1, p. 7–13.

Sheppard, Harriet. (1833). "Notes on Some of the Canadian Song Birds." Lecture, 16 February. *Transactions of the Literary and Historical Society of Quebec*, Vol II.

Shteir, A. & Cayouette, J. (2019). "Collecting with 'botanical friends': Four Women in Colonial Quebec and Newfoundland." *Scientia Canadensis*, 41 (1), 1-30.

Towsey, Mark. (2010). "Exporting the Scottish Enlightenment: the reading experience of a colonial governor's wife." In M.J. Munro-Landi, ed. *L'Écosse et ses doubles: ancien monde, nouveau monde*. Paris: L'Harmattan.

A singular specimen of the potato: Natural history at the Literary and Historical Society of Quebec, 1824 to 1840

Bernatchez, Ginette. (1981). "La Société littéraire et historique de Québec (the Literary and Historical Society of Quebec) 1824-1890." *Revue d'histoire de l'Amérique française* 179–92.

Blair, L. (2016). "Moose in Flames: The Story of the Literary and Historical Society of Quebec." In Blair, L.; Donovan P., & Fyson, D. *Iron Bars and Bookshelves: A history of the Morrin Centre*. Montreal: Baraka Books.

Chartrand, Luc, Raymond Duchesne & Yves Gingras. (2008). *Histoire des Sciences au Québec de la Nouvelle-France à nos jours. Montreal: Boréal*.

Jarrell R. A. (1977). "The rise and decline of science at Quebec 1824-1844." *Histoire Sociale* 77–91

Literary and Historical Society of Quebec. (1824-1924). *The Transactions* & other publications.

Villeneuve, René. (2008). *Lord Dalhousie: Patron and Collector*. Ottawa: National Gallery of Canada.

Whitelaw, Marjory, ed. (1978). *Dalhousie Journals*. Ottawa: Oberon.

The be-knighted collector: James MacPherson Le Moine

Andrès, B. (2004). "Roger Le Moine dans notre mémoire." *Les Cahiers des dix*, (58), 105–133. Retrieved from https://doi.org/10.7202/1008119ar

Blair, L. (2016). "Moose in Flames: The Story of the Literary and Historical Society of Quebec." In Blair, L.; Donovan P., & Fyson, D. *Iron Bars and Bookshelves: A History of the Morrin Centre*. Montreal: Baraka Books.

Duchesne, R. & Carle, P. (1990). "L'ordre des choses: cabinets et musées d'histoire naturelle au Québec (1824-1900)." *Revue d'histoire de l'Amérique française*, 44(1), 3–30. Retrieved from https://doi.org/10.7202/304861ar

Hébert, Y. (1997). "Des oiseaux et des hommes : l'ornithologie de 1535 à nos jours." *Cap-aux-Diamants*, (51), 28–32.

Kuntz, Harry. (2010). "Science culture in English-speaking Montreal, 1815-1842." PhD Thesis, McGill University.

Le Moine J. M. & Le Moine R. (2013). *Souvenirs et réminiscences / glimpses & reminiscences* de James McPherson Le Moine. Presses de l'Université Laval.

Rajotte, P. (1997). "Les associations littéraires au Québec (1870-1895): de la dépendance à l'autonomie." *Revue d'histoire de l'Amérique française*, 50(3), 375–400. Retreived from https://doi.org/10.7202/305571ar

III – DRAWING NATURE

Drawing dissected mollusc penises: Natural history artists

Aitken, Richard. (2008). *Botanical Riches: Stories of Botanical Exploration*. Aldershot, UK: Lund Humphries.

Campbell, P. (2002). "In the Physic Garden." *London Review of Books* [Online] vol. 24 no. 18 p. 20. Available from https://www.lrb.co.uk/v24/n18/peter-campbell/in-the-physic-garden.

Daston, L., & Peter Galison, P. (2007). *Objectivity*. New York: Zone Books.

Fowkes Tobin, Beth. (1999). *Picturing Imperial Power: Colonial Subjects in Eighteenth-Century British Painting*. North Carolina, USA: Duke University Press.

Gates, William. (1939). *An Aztec Herbal: The Classic Codex of 1552*. Baltimore: The Maya Society.

Holmes, Richard. (2009). *The Age of Wonder: How the Romantic Generation Discovered the Beauty and Terror of Science*. New York: Pantheon Books.

Linda Hall Library of Science, Engineering and Technology. (2005). "Women's work: Portraits of 12 scientific illustrators from the 17th to 21st century." Retrieved from https://womenswork.lindahall.org/.

Middleton, Dorothy. (2004). "North, Marianne." In *Oxford Dictionary of National Biography*. Oxford, UK: Oxford University Press.

Parrish, Susan Scott. (1999). "Drawn from Life: Science and Art in the Portrayal of the New World by Victoria Dickenson (review)." *University of Toronto Quarterly*, vol. 69 no. 1, pp. 187-189.

Parsons, Harriet. (2005). "British–Tahitian collaborative drawing strategies on Cook's Endeavour voyage." Chapter 8 in *Indigenous Intermediaries: New perspectives on exploration archives*. In Shino Konishi, Maria Nugent, Tiffany Shellam (eds). Acton (Aus): Anu Press.

Roos, Anna Marie Eleanor. (2019). *Martin Lister and his Remarkable Daughters: The Art of Science in the Seventeenth Century Oxford*. UK: Bodleian Library.

Russell, Roslyn. (2011). *The Business of Nature: John Gould and Australia*. Canberra: National Library of Australia.

The Badianus Manuscript: An Aztec Herbal. (1552). University of Virginia: Historical Collections at the Claude Moore Health Sciences Library.

Wahlquist, Calla. (2017). "The Pecking Order: How John Gould dined out on the Birds of Australia." In the *Guardian*, Dec 30. Retrieved from https://www.theguardian.com/environment/2017/dec/30/pecking-order-how-john-gould-dined-out-on-the-birds-of-australia

While her husband mapped the river: Natural history artists in Canada

Beetz, Jeannette and Henry. (1977). *La merveilleuse aventure de Johan Beetz*. Montreal: Leméac.

Blair, Louisa. (2016). "Moose in Flames: The Story of the Literary and Historical Society of Quebec." In Blair, L, Donovan P, and Fyson D, *Iron Bars and Bookshelves, A History of the Morrin Centre*. Quebec: Septentrion.

Creese M. and T.M. (2010). *Ladies in the Laboratory III: South African, Australian, New Zealand, and Canadian Women in Science: Nineteenth and Early Twentieth Centuries*. Lanhan, US: Scarecrow Press.

Dickenson, V. (1998). *Drawn from Life: Science and Art in the Portrayal of the New World*. Toronto: University of Toronto Press.

Eakins Peter R. and Eakins, Jean Sinnamon. (2003). "Dawson, Sir John William." In *Dictionary of Canadian Biography*, vol. 12, University of Toronto/Université Laval. Available at http://www.biographi.ca/en/bio/dawson_john_william_12E.html.

Gagnon, François-Marc. *Codex canadensis*. "About the manuscipt." Library and Archives Canada. Retrieved from https://web.archive.org/web/20150924055440/http://www.collectionscanada.gc.ca/codex/026014-1000-e.html

Library and Archives Canada. "Johan Beetz, naturaliste." Retrieved from http://epe.lac-bac.gc.ca/100/205/301/ic/cdc/johanbeetz/main.html.

Parr Traill, Catherine. (1839). *The Backwoods of Canada: being letters from the wife of an emigrant officer, illustrative of the domestic economy of British America*. London: C. Knight. Retrieved from https://archive.org/details/backwoodsofcanadootrai_1

Robert, K. (2008). "Women's Botanical Illustration in Canada: Its Gendered, Colonial and Garden Histories (1830-1930)." MA Thesis, Concordia University. Retrieved from http://spectrum.library.concordia.ca/976097/1/MR45323.pdf

Robert, Michel. (2013). "Mémoire présenté dans le cadre de la consultation publique sur le plan de conservation du site patrimoinial du Sillery." March 14.

Stacey, Robert. "Art Illustration." In *Canadian Encyclopedia*. July 2015. Retrieved from https://www.thecanadianencyclopedia.ca/en/article/art-illustration

Sugars, Cynthia. (2022). "10 Settler Botanists, Nature's Gentlemen." In Leis A (ed). (2023). *Women, Collecting and Cultures beyond Europe*. New York: Routledge. Retrieved from https://doi.org/10.4324/9781003230809

Swinton, George. (2008). "Inuit Art." In *The Canadian Encyclopedia*. Retrieved from https://www.thecanadianencyclopedia.ca/en/article/inuit-art

Thwaites, Ruben Gold. (1897). *The Jesuit relations and allied documents* [microform]: *travels and explorations of the Jesuit missionaries in New France, 1610-17*. Cleveland: Burrows. Retrieved from https://archive.org/details/cihm_07537.

Wilson, Harry. (2014). "How maps were used to politicize early Canada." In *Canadian Geographic*, Dec 24. Retrieved from https://www.canadiangeographic.ca/article/how-maps-were-used-politicize-early-canada

IV – EXPLAINING NATURE

Seasick on the *Beagle*: Origins of the Origin...

Browne, E. Janet. (1995). *Charles Darwin: Vol. 1 Voyaging*. London: Jonathan Cape.

Burkhardt, F. & Smith, eds. (1985). *The Correspondence of Charles Darwin. Vol. I: 1821-1836*. Cambridge, UK: Cambridge University Press.

Clark, Ronald. (2017). *The Survival of Charles Darwin: A Biography of a Man and an Idea*. Princeton: Princeton University Press.

Darwin Correspondence Project, "Letter to JS Henslow. Letter no. 171." Retrieved from http://www.darwinproject.ac.uk/DCP-LETT-171.

Darwin Correspondence Project, Darwin to Lyell, 1858. "Letter no. 2285." Retrieved from http://www.darwinproject.ac.uk/DCP-LETT-2285

Endersby, J. (2020), *Imperial Nature: Joseph Hooker and the Practices of Victorian Science*. Chicago: University of Chicago Press.

Thomson, Keith Stewart. (2014). "HMS Beagle, 1820-1870." *American Scientist* 102.3 (2014): 218. Retrieved from https://www.americanscientist.org/article/h-m-s-beagle-1820-1870.

Feel it struggling between one's fingers: Alfred Russel Wallace and the theories of evolution

"Beccaloni, George." (2008). The Alfred Russel Wallace Website. Retrieved from http://wallacefund.info/.

Conniff, R. (2011). *The Species Seekers: Heroes, Fools and the Mad Pursuit of Life on Earth*. New York: W. W. Norton.

Darwin Correspondence Project. (2022). "Alfred Russel Wallace." Cambridge: Cambridge University. Retrieved from https://www.darwinproject.ac.uk/alfred-russel-wallace.

Knapp, Sandra. (2013). *Footsteps in the Forest: Alfred Russel Wallace in the Amazon*. London, UK: Natural History Museum.

Quammen, D. (1996). *The Song of the Dodo: Island Biogeography in an Age of Extinction*. New York: Scribner.

Shapin, S. (2010). "The Darwin Show." *London Review of Books* [Online] vol. 32 no. 1 pp. 3-9. Retrieved from https://www.lrb.co.uk/v32/n01/steven-shapin/the-darwin-show.

Van Wyhe, J. (2013). "Dispelling The Darkness: Voyage in the Malay Archipelago and the Discovery of Evolution by Wallace and Darwin." *World Scientific*.

Wallace, A.R. (1905). *My life: A record of events and opinions*. London: Chapman and Hall.

Wallace, A.R. (1872). *The Malay Archipelago: The Land of the Orang-utan and the Bird of Paradise: a Narrative of Travel, with Studies of Man and Nature*. London: Macmillan & Co.

"I laughed ... til my sides were almost sore": How the theory of evolution by natural selection was received

Elshakry, M. (2014). *Reading Darwin in Arabic, 1860-1950*. Chicago: University of Chicago Press.

Numbers, Ronald L., Stenhouse, J. (2001). *Disseminating Darwinism: The role of place, race, religion, and gender*. Cambridge: Cambridge University Press.

Secord, James. (2000). *Victorian Sensation: The Extraordinary Publication, Reception and Secret Authorship of 'Vestiges of the Natural History of Creation.'* Chicago: University of Chicago Press.

Ziadat, A. (1986). *Western Science in the Arab World: The Impact of Darwinism, 1860-1930*. London: Macmillan.

A chaos of fallen rocks: Passionate opposition in Quebec

Berthold, Etienne. "Á la découverte d'une œuvre oubliée: l'action éducative et scientifique de l'abbé Léon Provancher à travers la Collection Léon-Provancher de l'Université Laval." *Material Culture Review / Revue de la culture matérielle*, [S.l.], Jan. 2003. ISSN 1927-9264. Retrieved from https://journals.lib.unb.ca/index.php/MCR/article/view/17952/22018

Chartrand, Luc, Raymond Duchesne, and Yves Gingras. (2008). *Histoire des sciences au Québec* : de la Nouvelle-France à nos jours. Montreal: Boréal. Chap 6. "Les naturalistes à l'heure de Darwin."

Davis, Wade. (2009). *The Wayfinders: Why Ancient Wisdom Matters in the Modern World.* Toronto: House of Anansi Press.

Duchesne, R. (1981). "La bibliothèque scientifique de l'abbé Léon Provancher." *Revue d'histoire de l'Amérique française*, 34 (4), 535–556. Retrieved from: https://doi.org/10.7202/303903ar

Jean-Marie Perron, "Provancher, Léon." In *Dictionary of Canadian Biography*, vol. 12, University of Toronto/Université Laval, 2003. Retrieved from http://www.biographi.ca/en/bio/provancher_leon_12E.html.

Peter R. Eakins and Jean Sinnamon Eakins, "Dawson, Sir John William." In *Dictionary of Canadian Biography*, vol. 12, University of Toronto/Université Laval, 2003. Retrieved from http://www.biographi.ca/en/bio/dawson_john_william_12E.html.

Zeller, Suzanne. (2001). "Environment, Culture and the Reception of Darwin in Canada, 1859-1909." In Numbers, R. L., & Stenhouse, J. (eds.). *Disseminating Darwinism: The role of place, race, religion, and gender.* Cambridge: Cambridge University Press.

V - MAPPING NATURE BY BOAT

No room for idlers: Captain Cook in Quebec

Beaglehole, J. C. (1992). *The life of Captain James Cook.* Stanford University Press.

British Library, Captain Cook Birthplace Museum, Northeast Museums Library and Archives Council. Website on Captain Cook. Retrieved from http://www.captcook-ne.co.uk/ccne/timeline/canada.htm

Glyndwr Williams, "Cook, James." In *Dictionary of Canadian Biography*, vol. 4, University of Toronto/Université Laval, 2003. Retrieved from http://www.biographi.ca/en/bio/cook_james_4E.html

"History of the Sailing Warship in the Marine Art." Retrieved from https://www.sailingwarship.com/james-cooks-chart-of-the-st-lawrence-to-quebec-by-order-of-vice-admiral-charles-saunders-1759.html

Lockett, Jerry. (2010). *Captain James Cook in Atlantic Canada: The Adventurer and Map Maker's Formative Years.* Halifax: Formac Publishing.

Rigby, N., van der Merwe, P. (2002). *Captain Cook in the Pacific.* London: National Maritime Museum.

Kletke, Glenn. (2014). "History Of An Idea About Tupaia's Chart." Captain Cook Society. Retrieved from https://www.captaincooksociety.com/home/detail/history-of-an-idea-about-tupaia-s-chart#7

Canada's Arctic Dogsbody: Captain Bernier, 1853–1934

d'Souza, P. (2002, June 1). "Captain Bernier lives on." *Nunatsiaq News*. Retrieved from https://nunatsiaq.com/stories/article/captain_bernier_lives_on/

Finnie, Richard S. (1974). "Farewell voyages: Bernier and the 'Arctic'. *The Beaver*, Summer issue.

Finnie, Richard S. (1974). "Joseph Elzéar Bernier (1852–1934)." Arctic Institute of North America. Retrieved from https://pubs.aina.ucalgary.ca//arctic/Arctic39-3-272.pdf

Harper, K. (2005, May 13). "Taissumani: May 16, 1910 – Captain Bernier and the alienation of Inuit land." *Nunatsiaq News*. Retrieved from https://nunatsiaq.com/stories/article/taissumani_may_16_1910_-_captain_bernier_and_the_alienation_of_inuit_land/

MacEachern, Alan. (2004). "Cool customer: the Arctic voyage of J. E. Bernier." *The Beaver*, Aug-Sept issue.

Osborne, S. (2013). *In the Shadow of the Pole: An Early History of Arctic Expeditions, 1871-1912*. Dundurn Press.

Saint-Pierre, M. (2009). *Joseph-Elzéar Bernier*. Montreal: Baraka Books.

Tremblay, A. (1921). *The Cruise of the Minnie Maud; Arctic Seas and Hudson Bay, 1910-11 and 1912-13*. Quebec: Arctic Exchange.

Nothing more human than a ship: The Canadian Arctic Expedition

"Northern People, Northern Knowledge: The story of the Canadian Arctic Expedition 1914-18." Retrieved from https://www.historymuseum.ca/cmc/exhibitions/hist/cae/indexe.html.

Andrews, John T. (2012) Review of *Stefansson, Dr Anderson and the Canadian Arctic Expedition, 1913–1918. A Story of Exploration, Science and Sovereignty* by Stuart E. Jenness. Gatineau, Quebec: Canadian Museum of Civilization, Mercury History Paper 56, 2011. 415 pp. In *Arctic, Antarctic, and Alpine Research*: v. 44, 2012.

Bartlett, Robert A. (Robert Abram) (2007). *The last voyage of the Karluk: shipwreck and rescue in the Arctic, 1875-1946*. St. John's: Flanker Press.

Canada, Department of Mines. Various dates, 1922 to 1944. *Report of the Canadian Arctic Expedition 1913-18*. Volumes 3 to 14. Ottawa: King's Printer. Retrieved from https://www.biodiversitylibrary.org/item/117515.

Cavell, J. (2013). Review of *Stefansson, Dr. Anderson and the Canadian Arctic Expedition, 1913–1918: A Story of Exploration, Science and Sovereignty* by Stuart E. Jenness. 2011. Gatineau, QC: Canadian Museum of Civilization. *Polar Record* 49 (251): e20 (2013). Cambridge University Press 2013.

Cook, S. (2017). "Filming the 'northern front': the motion pictures of the Canadian Arctic Expedition." *The Northern Review*, 427-455, 427–455. Retrieved from https://doi.org/10.22584/nr44.2017.019

Harper, Ken. (2008). "Ruth Makpii Ipalook: 1911-2008." *Nunatsiaq News*, July 17.

Henighan, Tom. (2009). *Vilhjalmur Stefansson: Arctic Traveller*. Toronto: Dundurn Press.

Kikkert, P. (2016). "Canadian Arctic Expedition." In *The Canadian Encyclopedia*. Retrieved from https://www.thecanadianencyclopedia.ca/en/article/canadian-arctic-expedition

Kusugak, Michael. (2019). "Our Land, Our Strength." *Canadian Geographic*, April 1.

McKinlay, William Laird. (1976). *Karluk: The great untold story of Arctic exploration*. London: Weidenfeld & Nicolson. (The book was republished in 1999 as *The Last Voyage of the Karluk: A Survivor's Memoir of Arctic Disaster*).

Niven, Jennifer. (2001). *The Ice Master*. London: Pan Books.

Roberts, H. H., Jenness, D. (1925). *Canadian Arctic Expedition, Southern party, 1913–16. Eskimo songs: songs of the Copper Eskimos*. Ottawa: F.A. Acland.

Watson, Anne. (2013). "Canada's Unsung Expedition." *Canadian Geographic*, Jan 1.

VI - EXHIBITING NATURE

Les simples curieux: The birth of the great museums

Arnold, Ken. (2016). *Cabinets for the Curious: Looking Back at Early English Museums*. Abingdon Oxon: Routledge Taylor and Francis Group.

Benedict, Barbara M. (2012). "Collecting Trouble: Sir Hans Sloane's Literary Reputation in Eighteenth-Century Britain." *Eighteenth-Century Life*, vol. 36 no. 2, pp. 111-142.

Bleichmar, Daniela. (2018). *Visual Voyages: Images of Latin American nature from Columbus to Darwin*. Yale University Press.

Daston, Lorraine and Park, Katharine. (2012). *Wonders and the Order of Nature 1150-1750*. New York Cambridge Mass: Zone Books; Distributed by the MIT Press.

Delbourgo, James. (2017). *Collecting the World: The Life and Curiosity of Hans Sloane*. London: Allen Lane, an imprint of Penguin Books.

Findlen Paula. (1996). *Possessing Nature: Museums, Collecting and Scientific Culture in Early Modern Italy*. Berkeley: University of California Press.

Fontes da Costa, Palmira. (2009). *The Singular and the Making of Knowledge at the Royal Society of London in the Eighteenth Century*. Newcastle-upon-Tyne: Cambridge Scholars Publishing.

Fortey, Richard. (2010). *Dry Storeroom No.1: The Secret Life of the Natural History Museum*. London: Harper Perennial.

Gates, Barbara T. (2007). "Why Victorian Natural History?" *Victorian Literature and Culture*, Vol. 35, No. 2 pp. 539-549.

Grice, G. (2015). *Cabinet of curiosities: Collecting and understanding the wonders of the natural world*. New York, N.Y.: Workman Publishing.

Hamilton, James. (2018). *The British Museum: Storehouse of civilizations*. London: Head of Zeus.

Jones, Jonathan. (2002). "Take me to your Dodo." The *Guardian* Wed 17 Jul 2002. Retrieved from https://www.theguardian.com/culture/2002/jul/17/artsfeatures.museums

LeBlond, Sylvio. (1979). "Le Dr James Douglas, de Québec, remonte le Nil en 1860-61." *Les Cahiers des dix*, (42), 101–123. Retrieved from https://doi.org/10.7202/1016240ar.

Lippincott, L., Blühm, A. (2005). *Fierce friends: Artists and animals, 1750-1900*. Van Gogh Museum, Amsterdam., & Carnegie Museum of Art. London: Merrell.

McGregor, Arthur. (1994). *Sir Hans Sloane: Collector Scientist Antiquary Founding Father of the British Museum*. London: Published for the Trustees of the British Museum by British Museum Press in association with Alistair McAlpine.

Meadow, Mark A. (2002). "Hans Jacob Fugger and the Origins of the Kunstkammer," in *Merchants and Marvels: Commerce, Science, and Art in Early Modern Europe*, edited by Pamela Smith, Paula Findlen. New York: Routledge.

Pomian, Krzysztof. (1991). *Collectors and Curiosities: Paris and Venice 1500-1800*. Cambridge U.K.; Cambridge Mass. USA: Polity Press; Basil Blackwell.

Schepelern, H. D. (1990). "The Museum Wormianum Reconstructed: A Note on the Illustration of 1655." *Journal of the History of Collections*, Volume 2, Issue 1, 1990, Pages 81–85.

Wilson, David M. (2002): *The British Museum: A history*. London: British Museum Press.

Otis and the safety elevator: World's fairs

Auerbach Jeffrey. (2022). *The Great Exhibition of 1851: A Nation on Display*. New Haven CT: Yale University Press.

Bricker, A B. (2022). "Ho Ho No! There arose such a clatter." *Literary Review of Canada*, December issue.

Cantor, Geoffrey. (2012). "Science, Providence and Progress at the Great Exhibition." *Isis*, 103:439–459.

Corbillé, S. F., Emmanuelle. (2020). "Les animaux dans les expositions universelles au xixe siècle: monstration, ordonnancement et requalification du vivant. Paris et Londres, 1851-1889." *Culture & Musées*, 35, 211-241.

Duchesne, Raymond, and Paul Carle. (1990). "L'ordre des choses: cabinets et musées d'histoire naturelle au Québec (1824-1900)." *Revue d'histoire de l'Amérique française* 44.1 : 3-30.

Farr, James (ed.). (2003). "Exhibitions: Empire and Industry." (2003) In, *Industrial Revolution in Europe 1750-1914*. Vol. 9. Detroit, Michigan: Gale, pp. 98-101.

Geppert Alexander C. T. (2010). *Fleeting cities: imperial expositions in fin-de-siècle Europe*. New York, NY: Palgrave Macmillan.

Groupe de recherche ACHAC. (2002). "Zoos humains : L'invention du sauvage" (exhibition program). Paris: ACHAC. Retrieved from https://www.achac.com/zoos-humains/wp-content/uploads/2015/04/Programme-Zoos-Humains-WEB.pdf

Hock, Beata. (2018). "The cultural politics of World Exhibitions and Biennials." Seminar at University of Leipzig, Global and European Studies Institute 2017-18.

Judd, Thomas W. (2013). "World's Fairs." In Thomas Riggs (ed.). *St. James Encyclopedia of Popular Culture*. 2nd Edition. Cengage Learning. Detroit: St. James Press. Retrieved from https://archive.org/details/stjamesencyclopeoooounse_m3jo/page/439

Macdonald, Sharon. (1997), *The Politics of Display: Museums, Science, Culture*. London: Routledge.

Qureshi, Sadiah. (2011). *Peoples on Parade: Exhibitions, Empire, and Anthropology in Nineteenth-Century Britain*. University of Chicago Press.

Vodden, C., & Frieday, L., & Block, N. (2017). "Geological Survey of Canada." In *The Canadian Encyclopedia*. Retrieved from https://www.thecanadianencyclopedia.ca/en/article/geological-survey-of-canada

Zeller, S. (2014). *Inventing Canada: Early Victorian Science and the Idea of a Transcontinental Nation*. Montreal: McGill-Queen's University Press.

The calf with two heads: The first natural science museums in Quebec

Blair, L., Donovan P., Fyson D. (2016) *Iron Bars and Bookshelves: A History of the Morrin Centre*. Montreal: Baraka Books.

Carle, P., Gagnon, P. & Metzener, M. (1992). "Florian Crête, c.s.v., et le Musée éducatif de l'Institut des Sourds-Muets (1882-1970): vers une nouvelle muséologie scientifique." *Scientia Canadensis*, 16(1), 60–75

Chartrand, L., Duchesne, R., & Gingras, Y. (2008). *Histoire des sciences au Québec: De la Nouvelle France à nos jours*. Montréal: Boréal.

Duchesne, Raymond. (2003) "Chasseur, Pierre." In *Dictionary of Canadian Biography*, vol. 7, University of Toronto/Université Laval. Retrieved from http://www.biographi.ca/en/bio/chasseur_pierre_7E.html.

Duchesne, Raymond, and Paul Carle. (1990). "L'ordre des choses: cabinets et musées d'histoire naturelle au Québec (1824-1900)." *Revue d'histoire de l'Amérique française*, 44.1: 3-30.

Duchesne, Raymond. "Magasin de curiosités ou musée scientifique? Le musée d'histoire naturelle de Pierre Chasseur à Québec (1824-1854)." *HSTC Bulletin*, vol. 7, no. 2, May 1983, p. 59–79.

Gagnon, Hervé. (1993). "Des animaux, des hommes et des choses. Les expositions au Bas-Canada dans la première moitié du XIXᵉ siècle." *Histoire sociale/Social History* 26.52: 317.

Harvey, Jocelyn and George Lammers. (2016). "Art Galleries and Museums." *The Canadian Encyclopedia*, 19 July 2016, Historica Canada. Retrieved from https://www.thecanadianencyclopedia.ca/en/article/art-galleries-and-museums.

Le Spectateur canadien (Montreal), 21 August 1824.

Pilarczyk, Ian. (2003). "'Justice in the Premises': Family and Violence and the Law in Montreal, 1825-1850." Thesis, McGill University, Montreal.

Sheets-Pyenson, Susan. (1988). *Cathedrals of Science: The Development of Colonial Natural History Museums During the Late Nineteenth Century*. Kingston: McGill-Queen's University Press.

Wonders, Karen. (1999). "A Wilderness Besieged." Part II. *The Beaver*. December.

Zeller, Susan. (2019). *Inventing Canada: Early Victorian Science and the Idea of a Transcontinental Nation*. Toronto: University of Toronto Press.

Sinking into the mud: The Geological Survey and the Canadian Museum of Nature

C. Gordon Winder. (2003). "Logan, Sir William Edmond." In *Dictionary of Canadian Biography*, vol. 10, University of Toronto/Université Laval. Retrieved from http://www.biographi.ca/en/bio/logan_william_edmond_10E.html.

Canadian Museum of Nature. History and Buildings (website). Retrieved from https://nature.ca/en/about-us/history-buildings and https://nature.ca/index.php?q=en/about-us/history-buildings/historical-timeline

Levere, Trevor Harvey and Richard A. Jarrell. (1974). *A Curious Field-Book: Science & Society in Canadian History*. Toronto: Oxford University Press.

Suzanne Zeller and Gale Avrith-Wakeam. (2003). "Dawson, George Mercer." In *Dictionary of Canadian Biography*, vol. 13, University of Toronto/Université Laval. Retrieved from http://www.biographi.ca/en/bio/dawson_george_mercer_13E.html.

"The History of the Geological Survey of Canada in 175 Objects." (2017). Ottawa: Government of Canada. Retrieved from https://science.gc.ca/site/science/en/educational-resources/history-geological-survey-canada-175-objects

Victoria Memorial Museum National Historic Site of Canada. (website) Retrieved from https://www.pc.gc.ca/apps/dfhd/page_nhs_eng.aspx?id=477

Visitor Services at the Canadian Museum of Nature. (2013). "The VMMB: Drama since 1912." Retrieved from https://natureguyandgal.wordpress.com/.

Vodden, Christy (1992). "No Stone Unturned: The First 150 years of the Geological Survey of Canada." Ottawa: Geological Survey of Canada. Retrieved from https://web.archive.org/web/20081207042708/http://gsc.nrcan.gc.ca/hist/150_e.php

Zeller, Suzanne. (2014). *Inventing Canada: Early Victorian Science and the Idea of a Transcontinental Nation*. Montreal: McGill-Queen's University Press.

Zeller, Suzanne. (2000). "The Colonial World as Geological Metaphor: Strata(gems) of Empire in Victorian Canada." Osiris, Vol. 15, in *Nature and Empire: Science and the Colonial Enterprise*, pp. 85-107 University of Chicago Press.

Slate pencils: William Dawson and the Redpath Museum in Montreal

Bronson, Susan. (1991). "The Design of the Peter Redpath Museum at Mcgill University: The Genesis, Expression, and Evolution of an Idea About Natural History." *SSAC Bulletin* 17. Retrieved from https://dalspace.library.dal.ca/bitstream/handle/10222/71340/vol17_3_60_76.pdf?sequence=1&isAllowed=y

Duchesne, Raymond, and Paul Carle. (1990). "L'ordre des choses: cabinets et musées d'histoire naturelle au Québec (1824-1900)." *Revue d'histoire de l'Amérique française* 44.1: 3-30

Falcon-Lang, H., & Calder, J. (2006). "Sir William Dawson (1820–1899): a very modern paleo-botanist." *Atlantic Geology*, 41(2). Retrieved from doi:https://doi.org/10.4138/181

McGill University (n.d.). "Tea and Fossils: History of the Peter Redpath Museum." McGill, Redpath Museum. E-Newsletter, ed. Hans Larsson. Retrieved from https://www.mcgill.ca/redpath/about/history.

Novacek, M.J. & Wheeler, Q.D. (1992). *Extinction and Phylogeny*. Columbia University Press.

Sheets-Pyenson, S. (1982). "Stones and Bones and Skeletons: the Origins and Early Development of the Peter Redpath Museum (1882-1912)." *McGill Journal of Education*, 17,1: 49- 62. Retrieved from http://mje.mcgill.ca/article/view/7441/5371.

Sheets-Pyenson, S. (1987). "Cathedrals of Science: The Development of Colonial Natural History Museums during the Late Nineteenth Century." *History of Science*, 25(3), 279–300

Sheets-Pyenson, S. (1996). *John William Dawson: faith, hope and science*. McGill-Queen's University Press, Montreal.

Zeller, Suzanne. (2000). "The Colonial World as Geological Metaphor: Strata(Gems) of Empire in Victorian Canada." *Osiris*, vol. 15, pp. 85–107.

Sentinel species: natural history and the future of nature

Adde, A., C. Casabona i Amat, M. J. Mazerolle, M. Darveau, S. G. Cumming, and R. B. O'Hara. (2021). "Integrated modeling of waterfowl distribution in western Canada using aerial survey and citizen science (eBird) data." *Ecosphere* 12(10): e03790.

Bakker, Karen. (2022). *How Digital Technology Is Bring us Closer to the Worlds of Animals and Plants*. Princeton & Oxford: Princeton University Press.

Birds Canada [website]: https://www.birdscanada.org/volunteer.jsp?lang=EN

Canadian Museum of Nature [website]. Retrieved from https://nature.ca/en/research-collections/collections/cryobank

CBC News, Ottawa. "Cryobank houses frozen Noah's Ark." Sep 18, 2018. Retrieved from https://www.cbc.ca/news/canada/ottawa/museum-of-nature-cryobank-dna-frozen-storage-1.4827883

Chion C, Turgeon S, Cantin G, Michaud R, Ménard N, Lesage V, et al. (2018). "A voluntary conservation agreement reduces the risks of lethal collisions between ships and whales in the St. Lawrence Estuary (Québec, Canada): From co-construction to monitoring compliance and assessing effectiveness." *PLoS ONE*, 13(9): e0202560.

Corniu, Marine. (2018). "Lumière sur l'histoire géologique." *Québec Science*, 08-06-2017. Retrieved fromhttps://www.quebecscience.qc.ca/sciences/lumiere-sur-lhistoire-geologique/

Corriveau, Jeanne. "Une campagne de dénombrement au secours des papillons monarques." *Le Devoir*, 24 July 2018. Retrieved from https://www.ledevoir.com/societe/science/533037/au-secours-des-papillons-monarques

Drolet, Bruno. (2022). Presentation at the 2022 Meeting of the First Nations of Quebec and Labrador Sustainable Development Institute. Retrieved from https://fnqlsdi.ca/wp-content/uploads/2022/10/14-SCF_20220914_ProjetPineshishPiyaasiis_FR.pdf

eBird science [website]: Retrieved from https://science.ebird.org/en.

Espace Pour la Vie Montréal (Biodome). (2018). Blog and List of publications in 2017-2018: Insectarium employees and Associate Researchers. Retrieved from http://espacepourlavie.ca/en/scientific-publications-1.

Fleury, Isabelle. (2018). "La première banque de biodiversité au Canada." Radio Canada, L'heure de pointe Toronto, 18 Septembre 2018. Retrieved from https://ici.radio-canada.ca/premiere/emissions/l-heure-de-pointe-toronto/segments/entrevue/87572/banque-cryogenique-nationale-biodiversite

Howe, Ken. (2013). *The Civic-Mindedness of Trees*. Hamilton ON Canada: Wolsak & Wynn.

Journey North: Tracking Migrations and Seasons (website). Retrieved from https://journey-north.org/

Jungblut, Anne D., F. Raymond, V. Mohit, A. Culley, C. Lovejoy, J. Corbeil, and W.F. Vincent. "From The British Arctic Expedition (1875-76) to the Present: Application Of Genomics To Identify Historical and Modern Microbiomes as Sentinels of Arctic Change." Arcticnet Annual Conference, Quebec City, 2017. Oral presentations. Retrieved from http://www.arcticnetmeetings.ca/index.php?url=13217.

Lamothe, Michel. (2023). *Luminescence Dating in the Natural Sciences*. Elsevier Science Publishing Co Inc.

Lapointe, André. (2018). "Le collectionneur du mois de juin." 17 December. Nouvelles, Institut québécois de la biodiversité. Retrieved from https://iqbio.qc.ca/category/nouvelles/

Miller, Giles. (2017). "New study shows the value of historical sediment collections - Curator of Micropalaeontology." Natural History Museum Blog, 17 January. Retrieved from https://naturalhistorymuseum.blog/2019/01/17/new-study-shows-the-value-of-historical-sediment-collections-curator-of-micropalaeontology/#more-12937.

Minelli, Alessandro. (2022). "The species before and after Linnaeus." In John S.Wilkins, Frank E. Zachos, Igor Ya. Pavlinov (eds). *Species Problems and Beyond: Contemporary Issues in Philosophy and Practice.* Boca Raton: CRC Press, Taylor & Francis Group.

Scott Polar Research Institute, University of Cambridge. (2022). "British Arctic Expedition, 1875-1876." Retrieved from https://www.freezeframe.ac.uk/resources/expeditions/arctic/british-arctic-expedition-1875-6/british-arctic-expedition-1875-6

Shoreline Cleanup [website]: https://www.shorelinecleanup.ca/

Strasser, B. J. (2019). *Collecting Experiments.* Chicago: University of Chicago Press.

Taylor, Bruce. (2011). *No End in Strangeness: New and Selected Poems.* 1st ed. Toronto: Cormorant Books.

Treu, G., Sinding, M.H.S., Czirják, G.Á., Dietz, R., Gräff, T., Krone, O., Marquard-Petersen, U., Mikkelsen, J.B., Schulz, R., Sonne, C. and Søndergaard, J. (2022). "An assessment of mercury and its dietary drivers in fur of Arctic wolves from Greenland and High Arctic Canada." *Science of The Total Environment*, p.156–171.

INDEX

CREDITS

p. 7, "The Two-Headed Calf" from *The Hocus Pocus of the Universe* by Laura Gilpin, copyright © 1977 by Laura Crafton Gilpin. Used by permission of Doubleday, an imprint of the Knopf Doubleday Publishing Group, a division of Penguin Random House LLC. All rights reserved. p. 13, Miriam P. Blair, 2019, Miriam P. Blair Collection; p. 15, Robin Kimmerer, 2011, "Goldenrod and Asters," The Natural Histories Project, https://natural historiesproject.org/conversations/goldenrod-and-asters; p. 15, Miriam P. Blair , 2019, Miriam P. Blair Collection; p. 17 & colour insert #1, Exit from Noah's Ark, Bedford Group, ca. 1420, Paris; British Library, London/ Wikimedia; p. 19 James Ferguson, 1756, "The planetary motions as seen from the earth" Plate III, p. 64, in *Astronomy Explained upon Sir Isaac Newton's Principles and Made Easy to Those Who Have Not Studied Mathematics*, p. 64. London, Longman/Wellcome Collection, Public Domain Mark; p 20 & cover, North American Owls, Alexander Wilson, 1835. "Edinburgh Journal of Natural History and of the Physical Sciences, with the Animal Kingdom of the Baron Cuvier," Vol. 1. Edinburgh [etc.], Published for the proprietor [etc.], 1835-1840; Ernest Mayer Library, Harvard/ Biodiversity Heritage Library; p 21, Jean Deshayes, De la grande rivière de Canada appellée par les Europeens de St. Laurens (detail). Paris: chez N. de Fer, 1715. BAnQ. Canada: Public Domain; p. 23, [Unknown], Kew Garden's flagpole felled by MacMillan Bloedel, Copper Canyon Division, 1958; University of British Columbia/Library/Rare Books and Special Collections/MacMillan Bloedel Limited fonds, RBSC-ARC-1343-BC-1930-166-2-3: p. 24, Jean-Victor Dupin, 1775-1778, in Buchoz, P-J. *Histoire universelle du règne végétal, ou nouveau dictionnaire physique et economique de toutes les plantes qui croissent sur la surface du globe.* Brunet, Paris. Peter H. Raven Library, Missouri Botanical Garden/Biodiversity Heritage Library; p. 26 & colour insert # 2, W. Hart, 1887. Plate 18 in *A monograph of the Trochilidae, or family of humming-birds* by John Gould. London, Henry Sothern & Co, 1887. Smithsonian Institution/ Biodiversity Heritage Library; p. 28, Mary Anning, 1823. Sketch of plesiosaurus in autographed letter re. the discovery of plesiosaurus; Wellcome Collection, Public Domain Mark; p. 30, J. Delarue, 1850, *Balanophyllia desmophyllum, Dendrophyllia dendrophylloides, Stephanophyllia discoides,* in "British fossil corals" by Henri Milne-Edwards & Jules Haime. London, Palaeontographical Society, 1850-54. p. T6. Smithsonian Libraries/Biodiversity Heritage Library; p. 32, Lafitau, P., 1724, in *Mœurs des sauvages americains comparées aux moeurs des premiers temps,* Vol. 1, Paris, Saugrain l'Âiné; Wikimedia/Public Domain; p. 34, Gordons, 1900-1930, Frank and Frances Carpenter Collection, Library of Congress, Washington, LC-USZ62-133503 (b&w film copy neg.); p. 36, unnamed Indian artists, 1819. In William Roxburgh's *Plants of the coast of Coromandel: selected from drawings and descriptions presented to the hon. court of directors of the East India Company*; Vol III (1819), Plate 259. Missouri Botanical Garden's Rare Books Collection/Biodiversity Heritage Library; p. 37, Sir John Watson Gordon, 1837. Sourced and reproduced with permission from a private collection; p. 39, [Unknown], ca. 2007, Collection LHSQ; p. 40, Young & Delleker, 1827, in Finley, Anthony, *A New General Altas, Comprising a Complete Set of Maps, representing the Grand Divisions of* he Globe, Together with the several Empires, Kingdoms and States in the World; Compiled from the Best Authorities, and corrected by the Most Recent Discoveries, Philadelphia. Geographicus Rare Antique Maps/Wikimedia; **p. 42**, Mawe, John, 1825, Instructions for the use of the blow-pipe, and chemical tests, with additions and observations derived from the recent publication of Professor Berzelius, frontispiece. Printed for and sold by the author ... and by Longman, Hurst, Rees, Orme, Brown, and Green, London, 4th edition. Wellcome Collection, Public Domain Mark; **p. 43**. L. Blair, 2013, copperplate of map by Alexander Sheriff, used for printing in the LHSQ Transactions, 1831. LHSQ Archives, BAnQ, ULaval (printed version: Local class no.: H3/400/1831, Box 2000216119, Item ID 4128134); **p. 45**, "JTW", 1887, in J.M. Le Moine, Chasses et Pêches au Canada, Quebec, N. S. Hardy, p. 148, library of the Literary and Historical Society of Quebec; **p. 46 & colour insert #5**, Elizabeth Gould, 1840, in Gould, J. The Birds of Australia, Vol III, Plate 14. London, [the author], Smithsonian Libraries/Biodiversity Heritage Library. **p. 48**, A & S Lister, 1685. In Historia Conchyliorum, by Lister, M. London: [The author].... Wellcome Collection, Public Domain Mark; **p. 49**, Blackwell, E. 1735, Plate 2 in A curious herbal, containing five hundred cuts, of the most useful plants. London: Printed for Samuel Harding ..., 1737, National Library of Medicine, Public Domain, NLM ID 101456747; **p. 51**, Julia Margaret Cameron, 1877, in Ford, C. Julia Margaret Cameron: 19th Century Photographer of Genius, ISBN 1855145065/ Wikimedia. Public Domain. **p. 52 & colour insert & 4**. Herman Moll, 1736. Inset from the so-called Beaver Map, A new and exact map of the dominions of the King of Great Britain on ye continent of North America : containing Newfoundland, New Scotland, New England, New York, New Jersey, Pensilvania, Maryland, Virginia and Carolina. New York Public Library Digital Collections. 1736.

p. 53, Louis Nicolas, ca. 1700. Inscription: "Licorne de La mer rouge ou l'on En voit quelques unes on en a porté ou conduit a medine, et a la meque ou les caravanes qui y vont En ont veu." In Codex canadensis, p. 27. Library and Archives Canada/ Wikimedia. **p. 54**, Agnes Fitzgibbon, 1869. Notes for colourist's issue for Traill, C.P., Canadian Wild Flowers. Montreal: John Lovell. McMaster Digital Collections. Original in Thomas Fisher Rare Book Library, University of Toronto, cham 00006, copy 2. **p. 55 & colour insert #6**, H. Gronvold, 1862. p. 413 in Notes on Anthropoid Apes by Walter Rothschild, Proceedings of the Royal Zoological Society, 1904. Smithsonian Libraries/Biodiversity Heritage Library; **p. 56**, R.T. Pritchett, based on drawing by Philip Gidley King, 1832. In Darwin, C. A Naturalist's Voyage Round the World, Journal of Researches into the Natural History and Geology of the countries visited during the voyage round the world of H.M.S. Beagle under the command of Captain Fitz Roy, R.N., London, J. Murray (1890 edition), p xiii. Smithsonian Libraries/Biodiversity Heritage Library; **p. 59**, John Gerrard Keulemans, from drawing by A. R. Wallace, 1869. In Wallace, A.R., The Malay Archipelago: The Land of the Orang-utan and the Bird of Paradise: a Narrative of Travel, with Studies of Man and Nature, Vol 1, p. 60. London: R. Clay Sons, & Taylor. Wellcome Library/ Biodiversity Heritage Library; **p. 62**, André Gill, 1878, Cover, La Petite Lune, no. 10. Bibliothèque nationale de France/Wikimedia; **p. 64**, Cover of 1st issue of Naturaliste canadien, 1869. (Seals picture taken from Frost, John, Grand illustrated encyclopedia of animated nature: embracing a full description of the different races of men, and of the characteristic habits and modes of life of the various beasts, birds, fishes, insects, reptiles, and microscopic animalcula of the globe. Being a complete history of the animal kingdom. Auburn, N.Y, The Auburn Publishing Company, E.G. Storke, Publishing Agent, 1855, p. 99.)/Smithsonian

Libraries/Biodiversity Heritage Library; **p. 65**, A.G. Ruggles, 1918. in Ruggles, A. G., *Report of the State Entomologist of Minnesota to the Governor*. St Paul, Minnesota: Agricutural Experiment Farm, p. 149/Biodiversity Heritage Library; **p. 67 and colour insert #7**, François Étienne Musin, 1846, National Maritime Museum, Greenwich, London, Caird Collection, BHC3325; **p. 69**, James Cook, 1759. Musée de la civilisation, collection du Séminaire de Québec, fonds Viger-Verreau; **p. 72**, [Unknown], 1905, J.E. Bernier/ Bibliothèque et archives Canada/C-001384; **p. 73**, [Unknown], 1924-1930, Library and Archives Canada, Fonds J-E Bernier, Album 8, Arctic, R216-998-2-E, Volume #14948, item # a102437-v8, 15/24; **p. 74**, [Unknown], 1900-1930, Frank and Frances Carpenter Collection, Library of Congress, Prints & Photographs Division, 11453-3, no. 42; **p. 76**, Lomen Bros, Nome, 1900-1930, Frank and Frances Carpenter Collection, Library of Congress, 11453-3, no. 16 [P&P]; **p. 77 & colour insert #8**, Cornelius Kreighoff, 1846, Gift of the Sigmund Samuel Endowment Fund, 1954. Royal Ontario Museum, 954.188.2; **p. 78**, Ole Worm, 1655, copper engraving print, Catalogue of Museum Wormiani Historia, Wellcome Collection, Public Domain Mark; **p. 80**, Jean-Baptiste Coriolan, 1642, in Aldrovandi, U. *Monstrorum Historia Cum Paralipomenis historiae omnium animalium*, Bologna, Typis Nicolai Tebaldini, p. 388, McGill University Library/ Wellcome Collection, Public Domain Mark; **p. 81**, Jean-Baptiste Coriolan, 1642, in Aldrovandi, U., *Monstrorum Historia Cum Paralipomenis historiae omnium animalium*, Bologna, Typis Nicolai Tebaldini, p. 388; McGill University Library/ Wellcome Collection, Public Domain Mark; **p. 82**, Joseph Paxton, 1851/Wikipedia; **p. 84**, [Unknown], 1924-1925, 1969-186 NPC, 4818, 279700, PA-181562. British Empire Exhibition Collection, National Archives of Canada; **p. 85**, Georges Bedon, 1900, L'Exposition de Paris (1900). In Encyclopédie du siècle. Exposition universelle 1900. Paris: Montgredien & Cie. vol. 2, p. 221; **p. 87**, James Smillie, engraver, 1831, Lande Collection, McGill University; **p. 88**, L. Blair, 2019, originally at Collège Bourget, Collections de l'Université Laval; **p. 89**, William Logan, 1844, Llyfrgell Genedlaethol Cymru – The National Library of Wales, NLW MS 21715-16B; **p. 90**, William Logan, 1845, McGill University Archives, Rare Books and Special Collections, Fonds MG 2046 - William Edmond Logan Fonds, NLW MS 21715-16B, f18r; **p 91**, William Logan, 1865, detail from *Geological Map of Canada and the Adjacent Regions, Including parts of other British Provinces and of the Univted States. The Geology of Canada, being derived from the Canadian Geological Survey; that of the other British Provinces from the labors of Dr. JW. Dawson, Professors James Robb, J. B. Jues and others*. Geographical Survey of Canada, doi:10.4095/133901. **p. 93**, [Unknown], 1911, Canadian Museum of History, 15276.5, CD95-812; **p. 94**, William Dawson, 1873, in Dawson, W. *The Story of Earth and Man*, New York, Harper & Brothers, 1873, page iv/Biodiversity Heritage Library; **p. 96**, Wm Notman & Son, 1925, McGill University, (1885-1956). McCord Collection, Purchase funds graciously donated by Maclean's magazine, the Maxwell Cummings Family Foundation and Empire-Universal Films Ltd, VIEW-2604; **p. 97**, The Tree of Life Explorer, Evogeneao, https://www.evogeneao.com/en/ explore/tree-of-life-explorer; **p. 98**, T. H Tizard (?), In Murray, John et al., *Report on the scientific results of the voyage of H.M.S. Challenger during the years 1873-76 under the command of Captain George S. Nares ... and the late Captain Frank Tourle Thomson, R.N.* Edinburgh, Neill, 1885, University of Toronto Gerstein Science Information Centre/Biodiversity Heritage Library; **p. 99**, Martin Fortier/Arctic Net, 2008, by kind permission; **p. 100**, Moses Griffith, 1806-1812, "Racoon ... American Pole Cat. ..." in Pennant, David [Untitled], original watercolour, wash, and pencil drawings of birds, animals, fishes, etc. V. 1., p. 67, Great Britain, Courtesy of The Linda Hall Library of Science, Engineering & Technology. **Colour insert #3**: E. Blackwell, E. 1735, Plate 2 in *A curious herbal, containing five hundred cuts, of the most useful plants.* London: Printed for Samuel Harding ..., 1737, National Library of Medicine, Public Domain, NLM ID 101456747.

MIX
Paper
FSC® C100212

Printed by Imprimerie Gauvin
Gatineau, Québec